Strategic Tillage and Soil Management - New Perspectives

Edited by Rodrigo Nogueira de Sousa

Published in London, United Kingdom

Strategic Tillage and Soil Management - New Perspectives
http://dx.doi.org/10.5772/intechopen.111208
Edited by Rodrigo Nogueira de Sousa

Contributors
Abhishek Singh, Adnan Zahid, Ahmad Raza, Aline Martineli Batista, Ammara Fatima, Asma Ayub, Christopher Mulielie, Daniel Mukhwana, David Kersting, Debabrata Ghoshal, Debashis Dutta, Debashish Dutta, Eboche Dave Khasakhala, Flora Ajwera, Gaus Azam, George Ayaga, John Manyengo, Kanchana Wickramarachchi, Lindani Ncube, Lílian Angélica Moreira, Mahima Dixit, Md. Shahinur Rahman, Michael Langemeier, Muhammad Tariq Manzoor, Mukhtar Ahmed, Nauman Shafqat, Omkar Singh, Patrick Mudavadi, Paulo Leonel Libardi, Pius Cheboi, Raphael Segun R. Bello, Rodrigo Nogueira de Sousa, Sajid Ali, Sanjeev Kumar, Shivangi, Simon Ogbeche Ogbeche Odey, Thaís Nascimento Pessoa, Uday Pratap Shahi, Vishnu D. Rajput

First published in London, United Kingdom, 2024 by IntechOpen
IntechOpen is the global imprint of INTECHOPEN LIMITED, registered in England and Wales, registration number: 11086078, 167-169 Great Portland Street, London, W1W 5PF, United Kingdom

British Library Cataloguing-in-Publication Data
A catalogue record for this book is available from the British Library

Additional hard and PDF copies can be obtained from orders@intechopen.com

Strategic Tillage and Soil Management - New Perspectives
Edited by Rodrigo Nogueira de Sousa
p. cm.

This title is part of the Agricultural Sciences Book Series, Volume 10
Topic: Agronomy and Horticulture
Series Editor: W. James Grichar
Topic Editor: Ibrahim Kahramanoglu
Associate Topic Editors: Murat Helvaci and Olga Panfilova

Print ISBN 978-1-83769-222-4
Online ISBN 978-1-83769-223-1
eBook (PDF) ISBN 978-1-83769-224-8
ISSN 3029-052X

For EU product safety concerns:
IN TECH d.o.o., Prolaz Marije Krucifikse Kozulić 3, 51000 Rijeka, Croatia,
info@intechopen.com or visit our website at intechopen.com.

IntechOpen Book Series

Agricultural Sciences

Volume 10

Aims and Scope of the Series

The importance of agriculture cannot be overstated. It helps sustain life, as it gives us the food we need to survive and provides opportunities for economic well-being. Agriculture helps people prosper around the world and combines the creativity, imagination, and skill involved in planting crops and raising animals with modern production methods and new technologies. This series includes two main topics: Agronomy and Horticulture, and Animal Farming. This series will help readers better understand the intricacies of production agriculture and provide the new knowledge that is required to be successful. The success of a farmer in modern agriculture requires knowledge of events happening locally as well as globally that impact input decisions and ultimately determine net profit.

Meet the Series Editor

W. James Grichar has been employed with Texas A&M AgriLife Research for over 45 years with an emphasis on research in agronomy, plant pathology, and weed science. He obtained his BS from Texas A&M in 1972 and his Masters of Plant Protection in 1975. He has published 195 journal articles, over 330 research reports and briefs, 11 book chapters, and over 300 abstracts of profession meetings. He also directs research in many crops including corn, grain sorghum, peanuts, and sesame. He has held various positions in different professional societies including the American Peanut Research and Education Society, Southern Weed Science Society, and Texas Plant Protection Conference in addition to being Associate Editor for Peanut Science and Weed Technology. Significant accomplishments have included spearheading efforts to determine the optimum planting time for soybean production along the upper Texas Gulf Coast. These efforts have shown growers that soybean yields can be improved by 10 to 20% by following a late March to early April plant date. He also has been instrumental in developing a herbicide program for peanut production in the south Texas growing region. Through the development and use of herbicides that are effective against major weed problems in the south Texas region, peanut yields have increased by 25 to 30%.

Meet the Volume Editor

Rodrigo de Sousa obtained a Bachelor's degree in agronomy engineering and a master's in soils and plant nutrition from the Federal University of Viçosa (UFV), Brazil, in 2016 and 2018, respectively. He was a scientific initiation scholarship holder throughout his undergraduate program in the Department of Soils/Stable Isotope Laboratory at UFV, focusing on organic matter and soil fertility. He participated in a sandwich undergraduate program at North Carolina A&T State University, USA, between 2014 and 2015. He completed an internship at the Department of Crop and Soil Sciences, North Carolina State University, USA, in 2015, working on nitrogen fertilizer management in maize production. Dr. de Sousa obtained a Ph.D. in the Soils and Plant Nutrition Program of Luiz de Queiroz College of Agriculture (ESALQ/USP), University of São Paulo, Brazil. He specializes in soil fertility/chemistry and plant nutrition.

Contents

Preface

Welcome to Strategic Tillage and Soil Management – New Perspectives, a comprehensive compilation dedicated to advancing knowledge and practices in agricultural soil management and tillage strategies. This preface introduces the breadth and depth of topics covered within this volume, serving as a guidepost for researchers, practitioners, and students who seek to enhance their understanding and application of soil science in agricultural contexts.

At the heart of this volume is a shared recognition of the critical role that soil plays in sustaining agriculture and, by extension, life on Earth. The chapters meticulously explore various facets of soil management, tillage, and plant nutrition each contributing unique insights into how we can protect, rehabilitate, and optimally utilize agricultural soils.

We begin with Chapter 1, "Plant Nutrition Optimization: Integrated Soil Management and Fertilization Practices" which provides a comprehensive and integral perspective on the interconnection between soil fertility and plant nutrition. Moreover, it offers insights into how strategic fertilization can significantly boost agricultural productivity.

Chapter 2, "Protection and Rehabilitation of Degraded Soils in Western Kenya: Experiences from Two Decades of Research and Development", sets the tone for the volume by addressing the urgent need for restoring soil health. Chapter 3, "Enhancing Agriculture through Strategic Tillage and Soil Management: Unleashing Potential for Sustainable Farming", expands on innovative approaches that promise to transform agricultural practices for the better.

As we delve further, Chapter 4, "Strategic Tillage for Improved Soil Health and Nutrient Dynamics", and Chapter 5, "Effective Microorganisms (EM): A Potential Pathway for Enhancing Soil Quality and Agricultural Sustainability in Africa", offer practical guidance on enhancing soil nutrient profiles and leveraging microorganisms for soil health, respectively. These chapters underscore the importance of integrating biological methods with strategic tillage to achieve sustainable soil management.

The exploration continues with Chapter 6, "Movement of Water and Solutes in Agricultural Soils", which provides a scientific foundation on the dynamics of water and nutrients within the soil, critical for informed irrigation and fertilization practices. Chapter 7, "Examining the Relationship between Crop Net Returns, Risk, and Conservation Practices", then bridges the gap between economic considerations and conservation efforts, illustrating the viability of sustainable practices from a financial standpoint.

In addressing the challenges and opportunities of tillage in specific climates, Chapter 8, "Strategic Tillage for Sustaining the Productivity of Broadacre Cropping in the Arid and Semi-Arid Regions of Southern Australia", highlights adaptive strategies for arid conditions. Chapter 9, "Modeling Growth and Yield of Crops Using Different Tillage Systems", further contributes by offering analytical models to predict crop outcomes under various tillage practices.

Finally, Chapter 10, "Harnessing Soil Potential: Innovation in Strategic Tillage and Management – New Perspectives", encapsulates the volume's theme, advocating for a forward-looking perspective on soil management and tillage innovation.

As we present this volume, we hope that it serves not only as a repository of current knowledge but also as a catalyst for further research, innovation, and application in the field of strategic tillage and soil management. May it inspire new perspectives and practices that contribute to the sustainability and productivity of agriculture worldwide.

Respectfully,

Rodrigo Nogueira de Sousa
Department of Soil Science,
Luiz de Queiroz College of Agriculture,
University of São Paulo,
São Paulo, Brazil

Chapter 1

Plant Nutrition Optimization: Integrated Soil Management and Fertilization Practices

Rodrigo Nogueira de Sousa and Lílian Angélica Moreira

Abstract

This chapter examines soil analysis, fertility management, and integrated soil management (ISM) practices that are critical to sustainable agriculture and environmental stewardship. It highlights the importance of detailed soil analysis—covering physical, chemical, and biological aspects—to inform decisions about fertilization, irrigation, and crop management. The discussion extends to soil nutrient dynamics, emphasizing how accurate analysis identifies imbalances and deficiencies that affect crops. ISM is presented as a holistic strategy that combines multiple dimensions of soil health to improve agricultural resilience. Practices such as conservation tillage, organic matter addition, and crop rotation are shown to increase soil fertility and achieve ecological goals, including biodiversity and carbon sequestration. In addition, the chapter advocates sustainable fertilization within ISM, promoting precision agriculture and 4R nutrient stewardship (right source, rate, time, place) to balance productivity with environmental stewardship. It aims to provide a comprehensive guide to modern soil management techniques that enhance productivity while ensuring soil health and sustainability, and emphasizes the integration of scientific knowledge with practical approaches to address global environmental and food security challenges.

Keywords: plant nutrition, soil analysis, integrated soil management, sustainable fertilization, environmental sustainability

1. Introduction

The complex relationship between soil health, plant nutrition, and agricultural productivity is a cornerstone of sustainable agriculture. As the world's population continues to increase, there is pressure on agricultural systems to produce more food without compromising the environment. This scenario requires a deeper understanding of plant nutrition, soil science, and their direct impact on plant growth and ecosystem health. The first part of this chapter lays the groundwork by exploring the complexity of soil composition and the factors that influence its fertility. Soil is not just a medium for plant growth; it is a dynamic system teeming with life, biological, and chemical reactions that play a central role in the health of plants and, by extension, the entire agricultural ecosystem.

The second part of the chapter focuses on the importance of soil analysis as a tool for maximizing agricultural yields while conserving natural resources. Soil analysis provides crucial insight into the soil's physical, chemical, and biological properties and provides a roadmap for effective soil management practices. By understanding the current state of the soil, farmers and agronomists can make informed decisions about irrigation, fertilization, and crop selection, all of which are critical to optimizing yields and maintaining long-term soil productivity.

Building on the knowledge gained from soil analysis, the chapter then moves into integrated soil management (ISM). ISM is a holistic approach that seeks to balance soil health's chemical, physical, and biological aspects. This section highlights how such an integrated approach improves soil fertility and contributes to broader environmental goals, including biodiversity conservation, carbon sequestration, and water quality improvement. Practices such as conservation tillage, organic matter incorporation, and efficient water management are explored as key components of ISM.

The fourth part of the chapter focuses on fertilization strategies within the integrated management framework. Adopting precise and sustainable fertilization techniques is crucial in modern agriculture. This section discusses the principles of "source, rate, time, and right place" in fertilizer application, emphasizing the need for a deep understanding of plant–soil interactions and the use of precision agriculture technologies. The integration of organic and inorganic fertilizers and a mindful approach to their environmental impact form a significant part of this discussion.

This chapter aims to provide a comprehensive guide to soil analysis and fertility management, integrated soil management practices, sustainable fertilization, and efficient nutrient use by plants strategies. By bridging the gap between scientific understanding and practical application, this chapter seeks to contribute to developing more sustainable and productive agricultural systems worldwide.

2. Soil analysis and fertility diagnosis

An essential part of agricultural management is soil analysis, critical to maximizing crop yields and protecting soil resources [1]. This practice is important since it provides an in-depth understanding of the soil' physical, biological, and chemical status. This data is essential for making informed decisions about irrigation, fertilization, and cultivation techniques, significantly improving agricultural productivity and environmental sustainability [2]. In addition, soil analysis can help detect contamination, excessive salinity, pH imbalances, and nutrient deficiencies, allowing growers to take proactive steps to correct these problems before they negatively affect crop development [3]. By understanding these factors, farmers can make informed decisions about fertilization, irrigation, and crop selection to maximize yields and maintain the long-term productivity of their land.

One of the key aspects of soil analysis is assessing the soil's available nutrient content [4]. This typically involves testing for essential elements such as nitrogen (N), phosphorus (P), potassium (K), (B), (Fe), and others, which deficiencies or imbalances can have a significant impact on crop growth and yield [5]. Analyzing soil pH is also critical as it affects nutrient availability and microbial activity.

Soils with an inappropriate pH hinder the uptake of essential nutrients by plants, leading to reduced yields and potential plant health problems [6]. Thus, understanding the soil's chemical attributes through analysis is crucial for making appropriate improvement decisions, such as liming to adjust pH levels.

DOI: http://dx.doi.org/10.5772/intechopen.114848

Soil texture is another important factor that can be assessed by soil analysis, as it results in different water retention capacities and drainage properties [7]. Understanding soil texture allows farmers to make informed decisions about irrigation and drainage management, which are essential to optimize crop growth and prevent waterlogging or drought stress. In addition to these factors, soil analysis provides insight into the overall health and fertility of the soil. This can include assessing organic matter content, cation exchange capacity, and the presence of any contaminants or pollutants. Farmers can implement soil management practices that promote long-term soil health and sustainability by assessing these aspects.

Effective soil analysis begins with representative sampling [8]. This involves selecting multiple points within a field, taking into account variations in topography, soil types, historical land use, and management practices [9].

Samples should be taken from different depths, depending on the purpose of the analysis, such as assessing surface fertility or subsoil conditions. The sampling technique must ensure that the collected samples represent the area of interest, avoiding atypical or contaminated areas such as those near roads or field margins. As shown in **Figure 1**, Sample 1 and Sample 2 differ due to their distinct management regimes [10], while Sample 3 is unique as it is derived from a garden area.

After collection, the soil needs to be analyzed, and the type of analysis depends on the owner's aims. The chemical analysis focuses on determining the presence of critical nutrients such as macronutrients, calcium (Ca), magnesium (Mg), phosphorus (P), potassium (K), sulfur (S), nitrogen (N), and micronutrients. Additional chemical tests may measure soil pH, cation exchange capacity (CEC), organic matter content, salinity, and others. Physical analysis evaluates aspects soil texture, structure, porosity, and bulk density, critical to understanding water retention and root penetration. Biological analysis evaluates soil microbial activity and biodiversity, including assessing specific soil enzymes such as acid phosphatase and beta-glucosidase. These

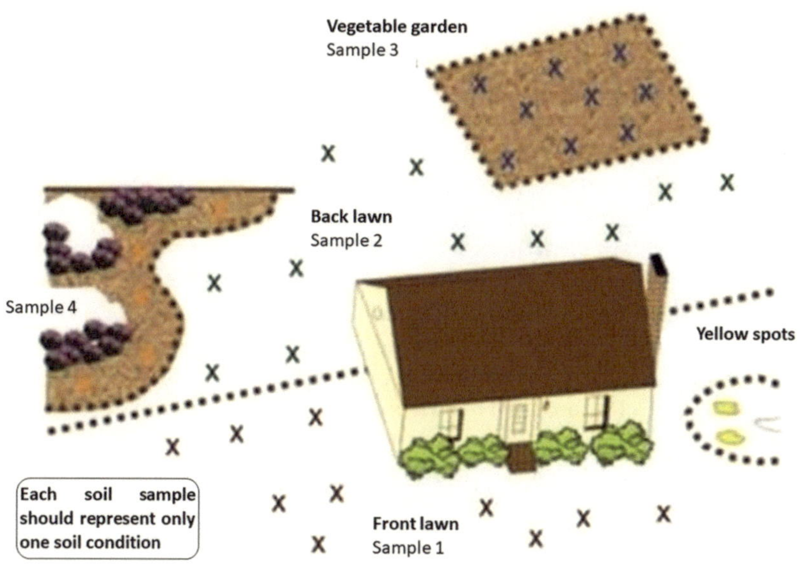

Figure 1.
Representation of different crops and management on the same farm, where soil sampling must be done individually within each system. Source: Ref. [10].

enzymes indicate nutrient cycling processes and overall soil health [11]. Together, these multiple analyses provide in-depth data on soil fertility and overall soil health, facilitating the development of field-specific management strategies.

The analysis results can first be used to interpret the actual condition of the soil at the time of sampling, such as soil acidity, compaction, nutrient deficiencies, and microbial activity. This diagnostic process begins with the interpretation of soil analysis results, where the concentration of macronutrients and micronutrients is compared to established agricultural benchmarks. These benchmarks are often region-specific, taking into account local soil types, climatic conditions, and crop requirements. The diagnosis helps to identify nutrient deficiencies, excesses, or imbalances, which is critical for developing an effective nutrient management plan [12]. The goal of soil fertility diagnostics to increase crop yields and minimize environmental impacts through the judicious use of fertilizers and soil amendments [13].

Soil nutrient depletion, particularly in tropical regions, is a major challenge to sustainable agriculture.

This place's soil is often characterized by low natural fertility due to intense leaching and high rates of organic decomposition accelerated by the hot and humid climate [14]. These factors result in rapid loss of nutrients, especially N and P (mainly by sorption process), which are essential elements for plant growth. A soil is considered nutrient-poor or low-fertility when its ability to supply adequate nutrients to crops is limited [15]. Low levels of organic matter, reduced cation exchange capacity (CEC), and low concentrations of essential macronutrients and micronutrients often indicate this [16]. Under these conditions, plants show signs of nutrient deficiencies, resulting in stunted growth, low yields, and, in severe cases, crop failure.

As mentioned before, each country or region develops its fertilizer recommendation guidelines tailored to local agricultural conditions and crop requirements. These guidelines are often compiled into bulletins or manuals that provide growers and agronomists with region-specific advice on optimal fertilizer application rates and practices.

3. Integrated soil management to improve fertility and plant nutrition

Integrated soil management (ISM) is a holistic approach that combines various agronomic practices to improve soil fertility and health, ultimately leading to sustainable agricultural productivity [17]. The core philosophy of ISM is to treat the soil not just as a medium for plant growth, but as a dynamic and living ecosystem [18]. This approach involves balancing the soil's chemical, physical, and biological aspects to create a more resilient and productive system.

Chemical balance is achieved by providing essential nutrients through organic and inorganic fertilizers while ensuring that soil pH is optimal for nutrient availability and plant growth [19]. Physical soil health is enhanced by conservation tillage, which maintains soil structure and reduces erosion, and by maintaining optimal soil moisture and aeration through proper irrigation and drainage practices [18].

The biological aspects of ISM focus on enhancing soil biodiversity and microbial activity, which are critical for nutrient cycling and decomposition of organic matter [20]. Incorporating organic matter into the soil, such as crop residues, compost, or green manures, stimulates biological activity and helps build soil structure. Cover crops play an important role in this management type, protecting the soil from

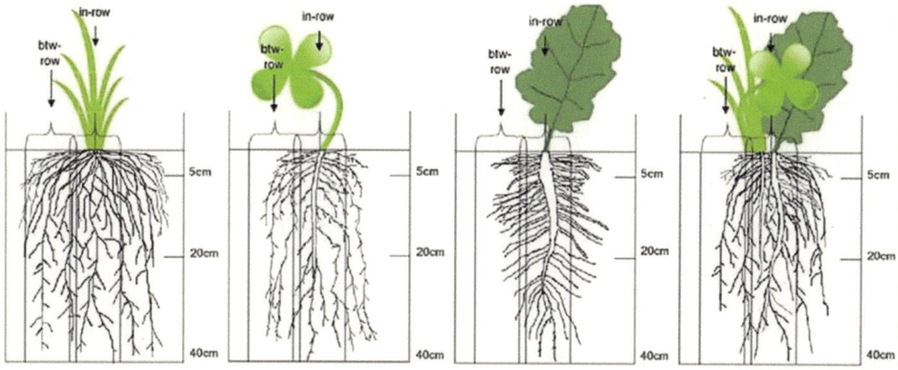

Figure 2.
Root distribution of triticale, crimson clover, canola, and a 5-species mix across In-row and between-row spaces at depths of 0–5 cm, 5–20 cm, and 20–40 cm. Source: Adapted from Amsili and Kaye [22].

erosion, adding organic matter and providing habitat for beneficial organisms [21]. As shown in **Figure 2**, according to Amsili and Kaye, some cover crop plants have root systems with different architectures and are generally quite robust, reaching deep into the soil [22]. In addition, crop rotation and intercropping practices support a diverse soil biota, which can improve nutrient availability, suppress soil-borne diseases, and increase overall soil resilience to environmental stresses.

Sustainability is at the heart of ISM, which means considering the long-term impact of soil management practices on the environment and future productivity [23]. This means adopting strategies that address current fertility issues, prevent soil degradation, and preserve soil health for future generations. Practices such as agroforestry, integrated pest and nutrient management, and conservation of natural soil habitats are integral to ISM [24]. These practices improve soil fertility and contribute to broader environmental goals such as biodiversity conservation, carbon sequestration, and water quality improvement. Ultimately, ISM is about striking a balance between meeting immediate agricultural needs and maintaining the ecological integrity of the soil ecosystem.

4. Fertilization strategies and integrated management

Fertility strategies and integrated management involve a multifaceted approach to nutrient application that ensures crops receive the nutrients they need for optimal growth while minimizing environmental impact [25]. The main focus of this approach is understanding soil nutrient dynamics and crop needs, which guides the selection and timing of fertilizer applications [26]. Modern fertilizer strategies emphasize the principle of "right source, right rate, right time, and right place," commonly known as 4R nutrient stewardship [27]. This involves selecting the appropriate type of fertilizer (organic or inorganic), applying it at the optimal rate to meet crop needs without over-application, timing the application to coincide with plant growth stages, and placing the fertilizer where the crop can most effectively utilize it. This targeted approach maximizes nutrient use efficiency and reduces the risk of runoff and leaching that can lead to environmental degradation [28].

Integrated fertilizer management also requires a deep understanding of crop–soil interactions. This includes recognizing the specific nutrient needs of different crops and the ability of different soil types to provide these nutrients [29]. Crop rotation and intercropping are critical to maintaining soil fertility and reducing reliance on chemical fertilizers. Rotating crops with different nutrient requirements can naturally help to balance the soil nutrient profile, intercropping with legumes can increase nitrogen availability through biological nitrogen fixation [30]. In addition, soil amendments such as lime can be used to adjust soil pH or improve soil structure, further optimizing nutrient availability [31].

The environmental impact of fertilization strategies includes the promotion of organic fertilizers and compost, which provide nutrients and improve soil organic matter content, soil structure, water retention, and microbial activity [32]. Integrating of organic and inorganic fertilizers can provide a balanced nutrient supply while maintaining soil health [33]. In addition, covering crops and reducing tillage practices as mentioned before contribute to soil health by helping to retain nutrients and reduce erosion, which is critical for sustainable nutrient management. Zhang et al. schematized an excellent example of integrated nutrient management as represented in **Figure 3** [34]. A strategic approach synergizes nutrient supply within the root zone with the nutrient needs of high-yielding crops to achieve sustainable agricultural productivity [35]. On the supply side, it quantifies bioavailable nutrients and characterizes the nitrogen (N), phosphorus (P), and potassium (K) available in the soil as influenced by environmental inputs. On the demand side, it considers the crop's total and periodic nutrient needs, as well as its response to nutrient inputs. The nutrient management strategy is twofold: It includes in-season management for nitrogen and addresses establishing and maintaining phosphorus and potassium levels. It also includes corrective actions for trace element deficiencies. These strategies are supported by high-yield crop management and optimal water management practices. The culmination of these practices forms the core of INM, which aims to balance crop productivity with environmental quality to ensure food security and long-term soil health.

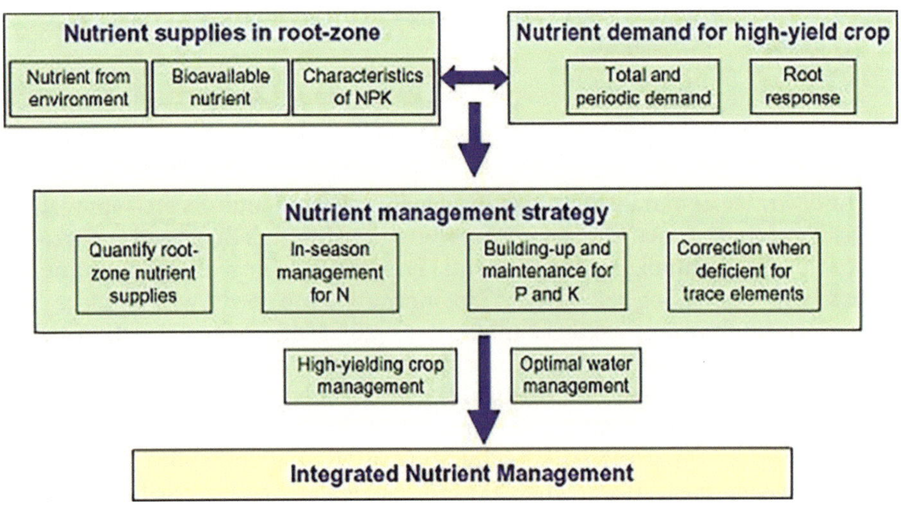

Figure 3.
Balancing soil health and productivity: Implementing integrated nutrient Management for Improved Food Security and Sustained Environmental Quality. Source: (adapted from Zhang et al. [34]).

Finally, continuous monitoring and adjustment are essential aspects of fertility strategies and integrated management. Soil fertility and crop nutrient needs can vary yearly due to changes in weather patterns, crop rotations, and other environmental factors [36]. Regular soil testing, plant tissue analysis, and plant growth and health monitoring are necessary to adjust fertilization strategies accordingly. This adaptive management approach ensures that fertilization practices remain effective and sustainable over time and meet the evolving needs of the soil–crop system. Furthermore, as mentioned before, an accurate fertilizer recommendation must be based on a detailed soil analysis and accurate interpretation of the results, taking into account crop history, regional climatic conditions, and specific crop requirements. Based on this information, agronomists can prescribe the exact type and amount of fertilizer to meet specific crop needs and identify soil deficiencies.

The selection of fertilizer sources is also critical. Different nutrient sources have different release rates and plant availability, affecting fertilizer efficiency [37]. For example, synthetic fertilizers provide a high concentration of nutrients and rapid availability, while organic fertilizers release nutrients more slowly but improve soil organic matter and structure [38]. The environmental impact of fertilizers should also be considered, with preference given to those with a lower risk of leaching or volatilization, such as controlled-release or stabilized fertilizers [37]. Precision agriculture technologies can further refine fertilizer recommendations, enabling site-specific and variable-rate applications that the crop's spatial and temporal needs [39]. As a result, informed fertilizer recommendations and wise selection of fertilizer sources are essential to successful integrated fertilizer strategies that positively impact agricultural productivity and natural resource conservation.

5. Use of nutrients by plants

Plant's efficient use of nutrients is essential for crop productivity and agricultural ecosystems' sustainability [40]. Plant nutrition is a delicate balance of nutrient uptake, assimilation, and utilization that must be tightly managed to avoid the depletion of soil resources and increase crop productivity [41]. Plants require a range of macronutrients and micronutrients, each serving a unique function within plant cells and systems. Macronutrients such as N, P, and K are often the focus of fertilizer programs because of their high plants demand and important role in growth and development. However, efficient use of these nutrients depends on their availability in forms that plants can readily absorb and utilize, which requires a nuanced understanding of soil chemistry and plant physiology [42].

The mechanism of nutrient uptake by plants is complex and involves soil–root interactions that are influenced by several factors such as soil pH, texture, and the presence of organic matter [43]. Mycorrhizal associations and root exudates modify the rhizosphere to enhance nutrient availability [44], and plant's root architecture can adapt to optimize nutrient uptake [43]. This biological efficiency is critical; for example, P, which tends to be fixed in the soil, can be made more available through the application of mycorrhizal inoculants or root organic acid exudates that solubilize phosphates and promote uptake [45].

Micronutrient efficiency is equally critical, although these nutrients are needed in smaller amounts. Micronutrients such as zinc, iron, manganese, and copper are essential for plant enzyme systems and photosynthesis [46]. Their efficient use is often hampered by soil conditions that make them unavailable to plants, such as high pH levels that cause these nutrients to precipitate. Chelated micronutrients or foliar applications can improve their availability and ensure that plants receive adequate amounts to maintain optimal physiological functions [47]. In addition, foliar application ensures a more uniform distribution, especially for nutrients required in low rates.

Water management also plays a critical role in nutrient use efficiency. Overirrigation can lead to leaching of nutrients, especially highly mobile ions such as nitrate, from the root zone, reducing their availability to plants and potentially contaminating groundwater [48]. Conversely, under-irrigation can result in poor solubilization of nutrients and reduced uptake due to reduced ion movement to the absorption zone [49]. Implementing precise irrigation strategies that deliver water directly to the root zone in amounts that meet plant needs without overdosing can significantly improve nutrient use efficiency [50].

6. Conclusions

This chapter highlights the importance of soil analysis and integrated management for sustainable agriculture. Soil analysis is essential for understanding nutrient levels, pH, and soil texture, which inform targeted fertilization strategies. Integrated soil management (ISM) combines chemical, physical, and biological approaches to improve soil health and fertility while protecting the environment through agroforestry and reduced tillage. Precision agriculture techniques are critical in optimizing fertilization to meet specific crop needs and protect the environment. Efficient nutrient use by plants is emphasized, focusing on the uptake and assimilation of both macro and micronutrients are crucial for plant growth. Ultimately, the chapter advocates a balanced approach that meets agricultural needs and maintains ecological integrity, ensuring long-term soil productivity and environmental sustainability.

Acknowledgements

The authors would like to thank the Coordenação de Aperfeiçoamento de Pessoal de Nível Superior—CAPES (financing code 001) for supporting the publication of this study. This research was supported by the Foundation for Research Support of the State of São Paulo, Brazil (FAPESP - 2020/13710-0), and the Brazilian National Council for Scientific and Technological Development (CNPq - 168192/2018-7; 371878/2023-3).

Conflict of interest

The authors declare that they have no conflicts of interest that could influence the publication of this chapter.

DOI: http://dx.doi.org/10.5772/intechopen.114848

Author details

Rodrigo Nogueira de Sousa[1*] and Lílian Angélica Moreira[2]

1 Department of Soil Science, Luiz de Queiroz College of Agriculture, University of São Paulo, Brazil

2 Laboratory of Stable Isotopes, Center for Nuclear Energy in Agriculture, University of São Paulo, Brazil

*Address all correspondence to: rodrigoagroufv@gmail.com

References

[1] Pande PK, Kumar R, Gupta S, Sachdev R, Mazumder MD, Satapathy L. Soil nutrient evaluation and crop management for sustainable growth of Patiya Village cluster in Almora, Uttarakhand. Current World Environment. 2021;**16**(3):818

[2] van Bavel CHM. Water relations of plants and soils. Soil Science. 1996;**161**(4):257-260. DOI: 10.1097/00010694-199604000-00007

[3] Ondrasek G, Rengel Z. Environmental salinization processes: Detection, implications & solutions. Science of the Total Environment. 2021;**754**:142432. DOI: 10.1016/j.scitotenv.2020.142432

[4] Wang J, Fu B, Qiu Y, Chen L. Analysis on soil nutrient characteristics for sustainable land use in Danangou catchment of the loess plateau, China. Catena. 2003;**54**(1-2):17-29. DOI: 10.1016/s0341-8162(03)00054-7

[5] Bindraban PS, Dimkpa C, Nagarajan L, Roy A, Rabbinge R. Revisiting fertilisers and fertilisation strategies for improved nutrient uptake by plants. Biology and Fertility of Soils. 2015;**51**(8):897-911. DOI: 10.1007/s00374-015-1039-7

[6] Fageria VD. Nutrient interactions in crop plants. Journal of Plant Nutrition. 2001;**24**(8):1269-1290. DOI: 10.1081/pln-100106981

[7] Osman KT. Physical properties of soil. In: Soils. Dordrecht: Springer; 2013. DOI: 10.1007/978-94-007-5663-2_5

[8] Khasawneh FE, Soileau JM. Soil preparation and sampling techniques for studying ion movement. Soil Science Society of America Journal. 1969;**33**(3):476-477. DOI: 10.2136/sssaj1969.03615995003300030039x

[9] James DW, Wells KL. Soil sample collection and handling: Technique based on source and degree of field variability. In: Soil Science Society of America. Madison; 2018. pp. 25-44. DOI: 10.2136/sssabookser3.3ed.c3

[10] McCray B. Soil Sampling and Tools - Oklahoma State University, Stillwater. 2020. Available from: https://agriculture.okstate.edu/departments-programs/plant-soil/soil- testing/soil-sampling-and-tools.html

[11] Neemisha P, Sharma S. Soil enzymes and their role in nutrient cycling. In: Giri B, Kapoor R, Wu QS, Varma A, editors. Structure and Functions of Pedosphere. Singapore: Springer; 2022. DOI: 10.1007/978-981-16-8770-9_8

[12] Pritts M, Heidenreich C, McDermott L, Miller J. Berry Soil and Nutrient Management–A Guide for Educators and Growers. Ithaca, NY: Cornell University; 2015

[13] Aleminew A, Alemayehu M. Soil fertility depletion and its management options under crop production perspectives in Ethiopia: A review. Agricultural Reviews. OF; 2020;**41**(2):91-105. DOI: 10.18805/ag.r-136

[14] Sanchez PA, Palm CA, Buol SW. Fertility capability soil classification: A tool to help assess soil quality in the tropics. Geoderma. 2003;**114**(3-4):157-185. DOI: 10.1016/s0016-7061(03)00040-5

[15] Schöl LV, Nieuwenhuis R. Soil Fertility Management. Wageningen: Agromisa Foundation; 2005

[16] Jatav HS et al. An overview of micronutrients: Prospects and implication in crop production. In: Aftab T, Hakeem KR, editors. Plant Micronutrients. Cham: Springer; 2020. DOI: 10.1007/978-3-030-49856-6_1

[17] kauffman, Sjef. Towards integrated soil management. ILEIA Newsletter. 1999;**15**(1/2):9-10

[18] Kibblewhite M, Ritz K, Swift M. Soil health in agricultural systems. Philosophical Transactions of the Royal Society B: Biological Sciences. 2007;**363**(1492):685-701

[19] Kumar R, Kumar R, Prakash O. Chapter-5 The Impact of Chemical Fertilizers on our Environment and Ecosystem. Vol. 35. New Delhi: AkiniNik; 2019. p. 69

[20] Thiele-Bruhn S, Bloem J, de Vries FT, Kalbitz K, Wagg C. Linking soil biodiversity and agricultural soil management. Current Opinion in Environmental Sustainability. 2012;**4**(5):523-528

[21] Das K, Biswakarma N, Zhiipao R, Kumar A, Ghasal PC, Pooniya V. Significance and management of green manures. In: Giri B, Varma A, editors. Soil Health. Soil Biology. Vol. 59. Cham: Springer; 2020

[22] Amsili JP, Kaye JP. Root traits of cover crops and carbon inputs in an organic grain rotation. Renewable Agriculture and Food Systems. 2021;**36**(2):182-191. DOI: 10.1017/S1742170520000216

[23] Stewart BA. Soil Management: Experimental Basis for Sustainability and Environmental Quality. Vol. 4. Boca Raton: CRC Press; 1995

[24] Mali M, Latha KM, Sidar S, Kiran S. Review of conservation agriculture practices for sustainable farming. Structure. 2023;**5**:6

[25] Wu W, Ma B. Integrated nutrient management (INM) for sustaining crop productivity and reducing environmental impact: A review. Science of the Total Environment. 2015;**512-513**:415-427

[26] Dungait JA, Cardenas LM, Blackwell MS, Wu L, Withers PJ, Chadwick DR, et al. Advances in the understanding of nutrient dynamics and management in UK agriculture. Science of the Total Environment. 2012;**434**:39-50

[27] Majumdar K, Johnston AM, Dutt S, Satyanarayana T, Roberts TL. Fertiliser best management practices. Indian Journal of Fertilisers. 2013;**14**:34-51

[28] Bruulsema T. Managing nutrients to mitigate soil pollution. Environmental Pollution. 2018;**243**:1602-1605. DOI: 10.1016/j.envpol.2018.09.132

[29] Gruhn P, Goletti F, Yudelman M. Integrated Nutrient Management, Soil Fertility, and Sustainable Agriculture: Current Issues and Future Challenges. Almadies: The International Food Policy Research Institute (IFPRI); 2000

[30] Liebman M, Dyck E. Crop rotation and intercropping strategies for weed management. Ecological Applications. 1993;**3**(1):92-122

[31] Bronick C, Lal R. Soil structure and management: A review. Geoderma. 2005;**124**(1-2):3-22. DOI: 10.1016/j.geoderma.2004.03.005

[32] Singh TB et al. Role of organic fertilizers in improving soil fertility. In: Naeem M, Ansari A, Gill S, editors. Contaminants in Agriculture. Cham: Springer; 2020. DOI: 10.1007/978-3-030-41552-5_3

[33] Yadav S, Lal M, Naresh RK, Yadav RB, Yadav AK, Yadav KG, et al. Effect of organic and inorganic nutrient sources on productivity, grain quality of rice and soil health in north-west IGP: A review. International Journal of Current Microbiology and Applied Sciences. 2019;**8**:2488-2514

[34] Zhang F, Cui Z, Chen X, Ju X, Shen J, Chen Q, et al. Integrated nutrient management for food security and environmental quality in China. Advances in Agronomy. 2012;**116**:1-40. DOI: 10.1016/b978-0-12-

[35] Sanginga N, Woomer PL, editors. Integrated Soil Fertility Management in Africa: Principles, Practices, and Developmental Process. Nairobi: CIAT; 2009

[36] Grant CA, Peterson GA, Campbell CA. Nutrient considerations for diversified cropping systems in the northern Great Plains. Agronomy Journal. 2002;**94**(2):186-198

[37] Timilsena YP, Adhikari R, Casey P, Muster T, Gill H, Adhikari B. Enhanced efficiency fertilizers: A review of formulation and nutrient release patterns. Journal of the Science of Food and Agriculture. 2014;**95**(6):1131-1142. DOI: 10.1002/jsfa.6812

[38] Shaji H, Chandran V, Mathew L. Organic fertilizers as a route to controlled release of nutrients. In: Controlled Release Fertilizers for Sustainable Agriculture. 2021. pp. 231-245. DOI: 10.1016/b978-0-12-819555-0.00013-3

[39] Shaheb MR, Sarker A, Shearer SA. Precision agriculture for sustainable soil and crop management. In: Soil Science-Emerging Technologies, Global Perspectives and Applications. London, UK: IntechOpen; 2022

[40] Fageria NK, Baligar VC, Li YC. The role of nutrient efficient plants in improving crop yields in the twenty first century. Journal of Plant Nutrition. 2008;**31**(6):1121-1157. DOI: 10.1080/01904160802116068

[41] Kathpalia R, Bhatla SC. Plant mineral nutrition. In: Plant Physiology, Development and Metabolism. Singapore: Springer; 2018. DOI: 10.1007/978-981-13-2023-1_2

[42] Chapin F III, Stuart. The mineral nutrition of wild plants. Annual Review of Ecology and Systematics. 1980;**11**(1):233-260

[43] Zhang, Fusuo S, Shen J, Zhu Y. Nutrient interactions in soil-plant systems. In: Lal R, editor. Encyclopedia of Soil Science. United Kingdom: Routledge; 2002. pp. 885-887

[44] Mimmo T, Pii Y, Valentinuzzi F, Astolfi S, Lehto N, Robinson B, et al. Nutrient availability in the rhizosphere: A review. Acta Horticulturae. 2018;**1217**:13-28. DOI: 10.17660/actahortic.2018.1217.2

[45] Jakobsen I, Leggett ME, Richardson AE. Rhizosphere microorganisms and plant phosphorus uptake. Phosphorus: Agriculture and the Environment. 2005;**46**:437-494

[46] Zewide I, Sherefu A. Review paper on effect of micronutrients for crop production. Journal of Nutrition and Food Processing. 2021;**4**(7):1-8

[47] Modaihsh AS. Foliar application of chelated and non-chelated metals for supplying micronutrients to wheat grown on calcareous soil. Experimental Agriculture. 1997;**33**(2):237-245

[48] Rashmi I, Shirale A, Kartikha KS, Shinogi KC, Meena BP, Kala S. Leaching

of plant nutrients from agricultural lands. In: Essential Plant Nutrients: Uptake, Use Efficiency, and Management. 2017. pp. 465-489

[49] Alam SM. Nutrient uptake by plants under stress conditions. Handbook of Plant and Crop Stress. 1999;**2**:285-313

[50] BAR-TAL, Asher et al. Practices that simultaneously optimize water and nutrient use efficiency: Israeli experiences in fertigation and irrigation with treated wastewater. Managing Water and Fertilizer for Sustainable Agricultural Intensification. Colombo: International Water Management Institute; 2015. p. 209

Chapter 2

Protection and Rehabilitation of Degraded Soils in Western Kenya: Experiences from Two Decades of Research and Development

Flora Ajwera, George Ayaga, David Kersting, Patrick Mudavadi, John Manyengo, Christopher Mulielie, Eboche Dave Khasakhala, Pius Cheboi and Daniel Mukhwana

Abstract

Over exploitation of soils in Western Kenya has resulted in nutrient depletion and soil erosion leading to low and inadequate food supply and incomes on many smallholder farms. This has been further exacerbated by climate change effects including unpredictable periods of drought and variable rainfall which is less frequent, reduced amounts but more in intensity. Best practices for sustainable land use are well known to farmers and are economically viable. Despite this, they are applied only to a limited extent. Disseminating these practices has been challenging, and quite often, soil protection does not receive the socio-political attention it requires. The responsible institutions often fail to tackle the issue effectively. As a result, farmers lack incentives, such as secure land rights for agricultural enterprises, to ensure that they utilize soil resources in a sustainable way. Smallholder farmers, who are the major agricultural producers, have inadequate know-how to apply soil-friendly farming methods on their farms, or the money for initial investments. This book chapter describes approaches that are being applied, assessed and promoted by the GIZ supported "ProSoil" in Western Kenya to explore the extent of soil degradation for sustainable soil protection and rehabilitation to provide a useful knowledge base to support decision makers in developing adoption strategies for sustainable crop production.

Keywords: degradation, nutrient depletion, rehabilitation, soils, Western Kenya

1. Introduction

Land is considered a valuable natural resource around the world and a key factor in production that provides essential agro-ecological services such as food, fodder, energy, and fiber for human life. However, over-exploitation of land for ecosystem services and income generation has resulted in extensive land degradation, persistent degradation of biological productivity and ecological integrity, and loss of human value [1, 2]. More than a fifth of the world's total land area is already degraded due to years of unsustainable land-use practices, land-use changes, and extreme weather events [3]. In some parts of the world, soil fertility, biodiversity, ecosystem services, and productivity have decreased significantly [4]. Recent studies have indicated that more than half of all fertile soils in the world are now classified as highly degraded, with 25–40 billion tons lost annually [5].

In Sub Saharan Africa (SSA), land degradation is estimated to affect about 67% of the total area of land (16.1 m km^2) with varying degrees of severity [6]. The main direct drivers contributing to land degradation in SSA include unsustainable agricultural practices, over-grazing of livestock, over exploitation of forests and woodlands [7]. In addition, a number of waterways in SSA have been subject to significant pressure, particularly in the semi-arid and sub-humid zones where rivers are over-exploited and eutrophication is prevalent [8]. Shiferaw et al. [9] attributed poor land and water management in SSH to the increasing population growth coupled with poor or inadequate incentive structures for local resource managers, climate change and rampart poverty.

In Kenya, recent studies show that 12–15% of the country's total land area, supporting 27% of its population, is suffering from severe to very severe degradation, with hotspots concentrated in the western and southern regions [10, 11]. Western Kenya is one of the most densely populated and impoverished regions in Kenya, with a population density rising to over 1046 people per KM^2 in Vihiga County which is above the national Kenyan average of 66 persons per Km^2 [12]. The region is characterized by low agricultural productivity coupled, with limited off-farm income options and with more than 58% of households living in absolute poverty [13].

In the past, Western Kenya's traditional land management was based on fallowing unproductive fields in order to restore soil fertility [14]. However, due to the high density of the rural population and land fragmentation, this practice is no longer sustainable. As a result, more people are looking for new land to farm, and they are moving into steep slopes and to other fragile agro ecosystems such as wetlands and adjacent forests all contributing to the increased rates of soil erosion, landslides, soil nutrient loss, soil acidity and physical degradation with the consequences of low productivity in the region [15].

The rate of landscape degradation is increasing at a rapid rate, with soils lost on average 0.5 t ha^{-1} $year^{-1}$ in 2017 compared to 0.3 t ha^{-1} $yr.^{-1}$ in 1995 [16]. Physical mechanisms of soil degradation prevalent in the region include erosion by water; compaction; hard setting; surface sealing; crusting; and degradation of soil structure [14]. The leading chemical processes include soil nutrient loss; soil organic matter loss; soil acidification; and dominant biotic processes such as pest infestations; changes in vegetation cover and composition (loss of habitat); changes in pollinator population and composition; and changes in soil biota [17, 18].

Nutrient mining is widespread on many smallholder farms in western Kenya, due to continuous cultivation, insufficient nutrient supplementation in relation to plant requirements, and high levels of land erosion, leaching, and removal of crop residue from the fields [19]. Fertile land that used to yield 2–4 t ha^{-1} cereal grains has become degraded, with cereal crop yields falling to less than one t ha^{-1} [20]. Although Western Kenya has good climatic conditions and high agricultural potential, the yields of the staple food crops; maize and beans have remained low, with an average of 1 ton of grain per hectare (1 t ha^{-1}) for maize and 2–5 t h^{-1} for beans with optimal land management

[1, 21]. Approximately 20 kg of NPK, 4 kg and 20.3 kg of S are extracted from the soil every season, without replacement, for every ton of corn grain produced [22].

Given the above scenario, further land degradation through poor land management should be avoided and, where degradation is already taking place, concerted action should be taken to stop and reverse it. This would help to accelerate the process towards a land degradation neutral future and the achievement of the 2030 Agenda for Sustainable Development (AFSD). The work reported in this chapter was largely based on "conservation and rehabilitation of degraded soils in western Kenya – ProSoil".

2. The analysis of barriers limiting the adoption of soil rehabilitation practices

Despite the reported success stories on the demonstration of improved soil rehabilitation practices mainly from the pilot locations, the level and rate of the adoption of the improved technologies still remain low leading to the current condition of land degradation [23]. The reasons for the low uptake of the available improved technologies by farmers who continue to use unsustainable land use practices is not well understood [24]. However serious concerns have been raised among the Government organizations both at national and county levels, research institutions and development partners on how to upscale the site based successful soil rehabilitation strategies to the landscape-level impact of the rehabilitation solutions.

There are no straightforward, definitive answers to the underlying constraints to the increased adoption of the tested and improved soil rehabilitation strategies by the smallholder farmers at landscape level. Numerous, intricate, and interrelated factors appear to hinder the processes hence contributing to the continued resource degradation, particularly land degradation, which threatens agricultural development. Coenen et al. [25] classified the key gaps, bottlenecks, and obstacles probably preventing the successful mainstreaming of rehabilitation measures on smallholder farms as; technological and knowledge, policy and institutional framework and socio-economic and financial.

2.1 Technological and knowledge barriers

2.1.1 Knowledge gaps on land degradation

Several years of research has generated adequate knowledge on soil degradation, particularly on the biophysical, chemical and biological processes in relation to their negative impacts on agricultural production [26]. This has provided the impetus to the understanding of land degradation patterns and trends as the baseline for the planning of sustainable soil management on farms. However the measurements of changes in land degradation remains unclear since it is often complicated as it occurs over both time and space. Moreover the estimation of economic losses caused by the destruction of the environment and critical ecosystems has not been fully quantified [27]. The absence of these crucial information on the underlying changes of land degradation and the inherent economic losses from the ecosystems both at the national and local levels has presented a challenge for decision-makers to visualize and evaluate the severity of the problem to necessitate the political good will required for the rehabilitation initiatives at landscape level.

2.1.2 Inadequate coordination and collaboration between stakeholders

Management of soil rehabilitation and biodiversity conservation can only be achieved if related conservation interventions are interlinked within target landscapes. However, many interventions for rehabilitating soils and conserving biodiversity in the past were carried out disjointedly by various organizations at the national and county levels. Frequently the activities and priorities for rehabilitation and conservation strategies are not unified, and in some cases, the coordinating bodies do not have sufficient representation and authority, or lack the capacity to operate. In addition, a large number of ministries are still involved in land and natural resource management, often lacking coordination, resulting in overlapping responsibilities [28].

As a result many of the soil improving technologies developed through research and past interventions have remained pivoted in small plots around which they were developed. These practices have not been extrapolated from pilot sites to the broader context, mainly due to weak or inadequate coordination, lack of awareness and low capacity among extension service providers on soil rehabilitation and conservation strategies [14]. Moreover the extension service providers are few and often constrained by inadequate resources to disseminate the improved technologies to the smallholder farmers for adoption to wider areas. The capacity of stakeholders to develop and disseminate soil improving technology is also limited. The linkage between farmers, extension and research is weak and in most cases does not address the specific challenges faced by farmers [14].

The prevalent compartmentalized approach among institutions and stakeholders to addressing land degradation has often masked the visualization of rehabilitation strategies at landscape level thus constraining the effective monitoring of the desired changes both in the long-term nationally and locally. As a result the effects of land degradation and biodiversity loss are not yet captured in the national data [14]. A more comprehensive ecosystem management approach that is more effective in connecting and intersecting the interactions between soil restoration and biodiversity conservation thus facilitating a holistic view of the impacts of rehabilitation strategies at scale would present a viable alternative strategy to rehabilitation and conservation in the region [29].

2.2 Policy and institutional barriers

2.2.1 Lack of policy harmonization

Comprehensive set of policies, strategies and action plans to address the issues of land degradation at national and county levels have been developed. Over 100 legal articles that directly regulate Kenya's agriculture industry have been documented. While some of these laws are redundant or in conflict with one another, others cannot be enforced, as is the case in other developing counties, due mainly to the lack of resources, both technical and financial, to tackle the issues of soil rehabilitation [30, 31]. Additionally there exits over 30 semi-autonomous national institutions with a wide range of mandates some of which are overlapping and in some cases contradictory. Overlapping and contradictory mandate result from the fragmentation and duplication of institutional missions and policy frameworks [30]. Users of land resources lack suitable channels for consultation as well as long-term security for crucial resources. This reduces security of tenure and motivation to invest on soil improvements that could raise yields in the long or medium term [32].

2.3 Socio-economic and financial barriers

2.3.1 The subsistence nature of production

Most smallholder farmers in developing countries including Kenya practice subsistence agriculture whereby crops are mainly grown to meet the needs of themselves and their families [33]. Reasons for this situation are many, however the major limitation for the smallholder farmers to invest in the enhancement of soils for increased and sustained production is attributed to poverty [14]. The smallholders are often forced to utilize short-term coping mechanisms in resource use rather than long-term investments in land and other sustainable resource practices. The constant demand on local resources brought on by the rapid population growth with little or no conservation strategies has accelerated the rate of vegetation loss and hence increased degradation [34].

Turning small-scale farming into a business case is the vision of many researchers and development partners, however, there are challenges that keep small-scale farmers from realizing this vision. Improving the profitability of sustainable production requires an incentive to make investments and use management practices that support land productivity rather than resource extraction (for short-term profits) or abandonment if the returns do not make business sense [14].

2.3.2 Inappropriate incentive structure, in particular land tenure arrangements

The slow pace of issuing land title deeds by the Kenya Government that would guarantee security of tenure has had a negative impact on land resources management. Land resource users often do not have sufficient long-term security for the resources on which they depend, nor do they have adequate consultation mechanisms. This limits their sense of ownership over the resource and their interest in investing in soil rehabilitation measures that will lead to increased yields in the medium or long term [35, 36]. As a result, the continued poor land use practices coupled with the lack of incentives to invest in soil rehabilitation practices has led to the current status of poor soil health which by extension has deprived the smallholder farmers from reaping the benefits of sustainable incomes and profit from soil resources.

2.3.3 Lack of value chain approach in production

Most smallholder farmers are not linked to any structured value chain. Their ability to increase productivity is constrained by supply chain inefficiencies due to low crop yields, limited quality storage capacity, which lead to high post-harvest losses; weak market linkages; poor safety standards; and lack of consumer trust in the quality of agricultural produce [37]. Moreover the smallholder farmers are largely confined to informal financial mechanisms and simple tools, such as local savings and loan groups, to meet their relatively basic financial service needs. Credit facilities to support smallholder initial investments are scarce and when available they are not effectively applied to attract private sector involvement [38]. In addition, little or no credit will normally be granted to individuals for the management of common resources, as compared to communities. Smallholder farmers also face disadvantages due to small plots, low levels of capital investments, a lack of bargaining power, and poor connections to the international market [39, 40].

2.3.4 Gender inequalities

Women in Kenya, as in most parts of Africa, make up 70% of all agricultural workers; 90% of hoeing and weeding work; 60% of all harvesting and marketing work; 80% of food preservation work; transport from farm to stores (80%), 90% of domestic labor for water and firewood; and nearly 100% of the food preparation work of households; however, they have little influence on decision making on the farm. Some gender-specific barriers to female farmers' productivity growth are insecurity in terms of limited access to land and other resources, low educational attainment, limited resources to purchase inputs, social limitations in meeting extension agents, and access to other sources of information. In addition, women face challenges in obtaining permits, grants, other services, and incidental work that directly benefit them [41].

Giz ProSoil has focused on addressing some of these challenges in the intervention process by taking a holistic and integrated approach to have the greatest impact in preventing land and ecosystem degradation, while enhancing soil conservation and restoration.

3. Pro soil stakeholder mapping and analysis

The many stakeholders who were engaged in the project's implementation were identified during the ProSoil preparation phase and are shown in **Table 1**. However, these stakeholders, had a range of abilities, interests and in some cases, rivalries. While the majority of them were involved in activities related to soil rehabilitation and ecosystem management, their approaches were often fragmented and uncoordinated. In order to establish a shared vision and strategy for the execution and up scaling of soil rehabilitation in Western Kenya, the ProSoil made an effort to bring these stakeholders together. The following categories were used to broadly group the identified parties and institutions.

3.1 Local communities

With a specific focus on women and youth groups, this group represents smallholder farmers and other small producer groups. The project involved Water Resource Users Associations (WRUAs) and constituted community-based self-help groups. This category was an essential project constituent, participating throughout the whole project lifecycle and being represented in all significant decision-making processes. Their main goal is to enhance household income and ensure food security, but they are constrained by the available resources and technology advancements.

3.2 County governments

The agriculture sector has been fully devolved to the county governments under the new constitutional dispensation, therefore any initiative in the sector requires

No.	Stakeholder	Interest	Capacities	
			Strengths	**Limitation**
1	Local farmers	secure livelihoods Increased income Security of tenure	land owners local knowledge labour force Existing Community based organizations	High poverty levels Limited technical knowledge Limited incentives for soil rehabilitation
2	Women and Youth	Food security Gainful income from labour	Labour force Experience Existing Organized groups	Empowerment to make decisions Lack of direct income Suppressive culture and traditions, inadequate skills and knowledge, limited capacities, low literacy among women, attitude and perception
3	National and County governments	Legislation, Capacity development, Resource mobilization, Food security Economic growth Sustainable development	County government Devolution of agricultural sector Financial resources Political will	Limited human resource Policy and institutional gaps Limited knowledge management Lack of comprehensive land use plan
4	KALRO	To contribute to increased productivity, commercialization and competitiveness of the agricultural sector through generation and promotion of knowledge, information and technologies that respond to client's demands and opportunities.	Human resource Research infrastructure Knowledge management	limited capacity to research on modern technologies Weak linkages between research, farmers/ communities and extension
5	WRMA	*To effectively regulate and manage water resources in collaboration with stakeholders for sustainable development*	Constitutional mandate on water resource management	Conflicting and overlapping roles and responsibilities with regulatory bodies Poor or lack of hydrological data quality for effective water resources planning and protection
6	KEFRI	To conduct research and provide information and technologies for sustainable development of forest and allied natural resources.	Human resource Research infrastructure Knowledge management	Weak linkages between research and extension Limited awareness about KEFRI mandate
7	Local NGOs	Support to rural development Advocacy of community rights Resource mobilization	Community mobilization Project management Lobby and advocacy	Limited resources Many NGOS with no coordination mechanism, Many CBOs with limited capacity

No.	Stakeholder	Interest	Capacities	
			Strengths	Limitation
9	Private sector	Increased business opportunities in soil rehabilitation initiatives Public private partnerships	Business oriented Financial resources Market linkages	Limited information of soil rehabilitation practices that could generate business

Table 1.
Stakeholder analysis.

their approval. Additionally, the established County Agricultural and Environmental Committees, which are in charge of managing all countywide environmental conservation-related concerns. It is the responsibility of county governments to enact and uphold legislation that encourage soil rehabilitation. They are also a major source of co-funding because to their continuous initiatives in the agricultural and resource sectors, making them crucial project stakeholders. The county governments have a vested interest in supporting sustainable economic growth and ensuring the livelihoods of their residents. However, the human and intuitive capacities of county governments to carry out their duty are limited [42].

3.3 The national government institutions

The National Government, through the Statutory and Regulatory Authorities, in particular the Ministry of Agriculture (MoA), the National Environmental Management Authority (NEMA), and the Water Resource Management Authority (WRMA), were key stakeholders based on their mandates in the sustainable management of the environment and natural resources.

3.4 Institutions for national and international research

The main local partner with extensive expertise research was the Kenya Agricultural and Livestock Research Organization (KALRO) that participate in agricultural-related research and development in the region in partnership with; Masinde Muliro University of Science and Technology (MMUST), Maseno University, and Jaramogi Oginga Odinga University, as well as the Kenya Forestry Research Institute (KEFRI). The problem has been getting the knowledge and technologies that these institutions have generated to the local small-scale farmers. There are important lessons to be learned from the research investments made by a several research collaborators including the International Centre for Insect Physiology and Ecology (ICIPE), World Agro-forestry Centre (ICRAF), Vi-Agroforestry and the International Centre for Tropical Agriculture (CIAT).

3.5 Non-government entities

A number of regional NGOS and CBOs are active in the project area and concentrate on a variety of interests. Based on their technical capabilities and shown success in community mobilization, training, and advocacy, the active local NGOs and CBOs working in the area were chosen and included in project execution.

3.6 Commercial sector

The project supported Public Private Partnership (PPP) activities to aid small producer groups in accessing inputs, processing their products, and marketing. As a result, local artisans who fabricate conservation Agriculture (CA) tools and equipment through Agriculture Technology Development Centers (ATDCs) were able to develop their entrepreneurial abilities, which will help to increase the sustainability of the project activities. Through the PPPs, the business sector was also given the option to provide input into the process as necessary and to take part in supporting the project operations (**Table 1**).

4. Intervention strategy as a paradigm shift (alternative)

The alternate strategies used by GIZ ProSoil to solve the problems posed by soil degradation and rehabilitation in Western Kenya are described in this section. The ideas are by no means novel, but they were creatively used to approach related issues in a fresh way.

4.1 An integrated approach to multi-stakeholder participation

The GIZ ProSoil used an integrated multi-stakeholder participation approach in conjunction with appropriate rehabilitation techniques to address the severe soil deterioration in Western Kenya. This strategy is increasingly being recognized as a key framework for solving intricate and connected agricultural and environmental problems. It brings together various stakeholders who often have competing interests but share a same landscape and offers creative ways to advance agriculture, enhance rural livelihoods, and coordinate across sectors.

How to move from successful but often fragmented and localized soil rehabilitation initiatives to entire landscapes that are fully integrated into sector and county development plans and budgets for sustainability was key challenge that GIZ ProSoil is tried to address. The goal of the intervention is to transfer knowledge gained from pilot sites of discrete programs to larger production areas and to improve coordination, bringing soil rehabilitation towards a more productive approach. The overall intervention strategy is anchored on four pillars: (1) Capacity building of stakeholders to embrace and adopt landscape targeted soil rehabilitation strategies, (2) Strengthening farmer links to structured agricultural input and output markets; (3) Supporting enabling policy environment stakeholders and institutional involvement and (4) Support to Knowledge Communication Practices.

4.2 Capacity building of farmer groups in soil rehabilitation practices

An initial capacity needs assessment (CAN) for a cross section of stakeholders including farmer groups, extension service providers, and input providers was conducted to identify vital capacity gaps that needed to be filled to support the project's successful implementation. In order to reach more farmers in a wider context, the initiative employed tried-and-true methods of agricultural extension and training, such as demonstrations, field days, agricultural fairs, and farmer-to-farmer extension. The project, in particular, engaged in trainer training i.e. community resources persons (CRPs), frontline agricultural extension officers and lead farmers.

Figure 1.
Mucuna cover crop (velvet bean). Source: Daniel GIZ ProSoil.

Common soil rehabilitation measures focused on Integrated Soil Fertility Management (ISFM) among others; maize-bean intercropping, push-pull technology, agroforestry, conservation agriculture (**Figure 1**), soil and water conservation measures, good agronomic practices and crop-animal combinations with positive results. ISFM is a comprehensive approach to the study of soil fertility that includes the full range of determinants and consequences; biological, physical, chemical, social, economic and political, land degradation. The strategic use of fertilizers as well as organic sources of nutrients to ensure efficient fertilizer use and crop productivity at the farm level are the fundamental principles of ISFM [43].

The project adopted participatory learning methods which incorporate experiential learning to help farmers learn about current issues in an informal setting within their own environment. Learning focused on organized groups of farmers and supported by trained lead farmers and frontline extension workers.

The project also supported research and extension service providers to build capacity to better serve farmers. In particular, the project trained frontline agricultural extension officers, including lead farmers, and equipped them with training manuals on soil improvement practices and other related training materials.

4.3 Strengthening linkages of farmers with agricultural input and output markets

Smaller production groups' better access to input and output markets will depend on how well organized and informed they are about the dynamics of the value chain. To this end, ProSoil has strengthened production groups through continuous capacity building in group dynamics, governance, and organizational/leadership skills. The project also created strong links and partnerships with buyers and streamlines consolidation at collection centers with groups. In addition, a stakeholder forum including buyers, farmers, agricultural agents, local entrepreneurs and other actors in the value chain was established to facilitate sharing of market information and integrated production and marketing. For the input market, the project connected groups of trained farmers with agricultural agents working in western Kenya.

4.4 Supporting policy development and institutional frameworks at the local level

Barrier analysis by ProSoil identified a number of policy and institutional barriers that hinder the scaling up of soil rehabilitation activities. These policy barriers include fragmented and overlapping institutional mandates as well as limited incentives for smallholder farmers to invest in soil rehabilitation activities.

To overcome these obstacles, the project considered a number of related policies and how the county governments could create a harmonized framework. The findings and recommendations of the policy review were used to develop national agricultural land management policy in collaboration with national governments. County governments, both executive and legislative, have been sensitized to domesticate national policies at county-level as framework for carrying out soil rehabilitation activities. To date, the counties that have domesticated the national agriculture soil management policy are now at different levels of implementation.

4.5 Knowledge management and communication

Despite a fairly complete knowledge and technology base on soil rehabilitation activities, the information is scattered across different institutions and in different formats so it is not easily accessible. The varying successes of soil rehabilitation measures tested at different demonstration sites by different parties, even within the same area, have not been fully documented or shared. The base scenario would see a continuation of this approach with less synergies and additions and thus result in duplicate efforts.

Knowledge management is a process that involves more than just the creation and exchange of data or information; it also calls for mechanisms that foster change in the perceptions of the individuals involved and the co-creation of new knowledge with the involvement of numerous people and organizations. Maintaining a "repository" of both technical and local knowledge is part of knowledge management. It calls for efficient and sustainable methods of gaining knowledge [44].

The project supported knowledge management for soil rehabilitation through a number of strategies. Firstly, the project documented past and ongoing best practices in soil rehabilitation and related research conducted in the area with the aim of harnessing knowledge in a 'reservoir. Information is stored both manually and electronically in formats that are easy to retrieve and share. Additionally, the project developed awareness materials, including posters, brochures, and conference materials. ProSoil has built stakeholder and knowledge management capacity and helped county government's sensitive to the need to develop a central repository for all land conservation and restoration stakeholders.

In addition, the project developed a transition strategy to ensure that project information and progress reports are disseminated to key stakeholders. The project works with a wide range of stakeholders at the county, national and global levels. Therefore, a good communication strategy is needed to ensure that all stakeholders are informed about the progress of the project in real time [45]. Communication strategy is a two-way channel for announcements and feedback from stakeholders. It also involves the use of multimedia, including local radio stations and audio-visual documentaries. Information has been disseminated locally, nationally and internationally through various channels including peer reviewed scientific journal papers, conferences and seminars' literature.

4.6 E-extension

Due to weak extension services delivery by government as a result of constrained human and financial resources the project adopted alternative extension service delivery approaches.

4.6.1 Integrated electronic extension method

The project has promoted and supported the use of relevant ICT tools like mobile phones, which are suitable for farmers and traders to have real-time access to information and markets for their produce. The promoted tools include; the KALRO GAPs APP, which provides real time good agronomic practices for common agricultural value chains, Plant Clinics developed in collaboration with CABI and KALRO to provide *insitu* recommendations on pest and disease management for various crops value chains. Additionally, the project promoted SMS and USSD and mobile services for disseminating agricultural information including weather to rural farmers. This promotion was done in partnership with mobile phone service providers such as SAFARICOM and Airtel.

4.6.2 Video expansion model

The ProSoil took advantage of the existing E-Extension platforms and promoted the use of E-Extension to youth, women and male farmers to enhance extension service delivery. It also improved the existing E-Extension platforms linking it to farmers in the region. E-Extension information is delivered to farmers in the local languages through easy-to-use downloadable extension videos that show new farming methods on their smartphones. E-Extension enables users to engage in personal and informal virtual communications and networks via social networking sites. In addition to social use, it has also been used to disseminate agricultural extension and consulting information and feedback. With the rise in smartphone availability, there has been an increase in the use of social media such as Facebook, Twitter, Skype, You Tube and WhatsApp.

4.6.3 Multimedia extension

Multimedia extension model was used to create awareness among stakeholders in Western Kenya. A number of radio and TV stations including West FM, Nyota, Citizen TV (Shamba Shape up), Mulemebe, and Nam Lolwe FM partnered with the project in the dissemination and awareness creation of sustainable soil rehabilitation practices.

4.7 Towards increased mechanization for smallholder farmers

Different tillage and soil conservation technologies were introduced *sequentially* to farmers, first on a small trial scale and then on a larger scale over larger plots of land. This step-by-step and innovative approach is considered more suitable for farmers' understanding and subsequent adoption and modification of tools and equipment. Farmers tested the tools on small plots of land before using them on larger plots. After observing the results, the engineers modified the tools to be more suitable and friendly to the farmer's situation. Three types of tillage systems are practiced in western Kenya: no-till or minimal tillage, row tillage and cover tillage. Row tillage

Figure 2.
Chaka Jembe. Source: GIZ ProSoil.

works well on fairly flat but poorly drained soil, while mulch is done on soil with a high capacity to retain waste. Tillage includes all the physical, mechanical, chemical or biological activities performed to prepare the seedbed for seed germination, germination and seed formation, root development and plant growth.

Chaka Jembe (**Figure 2**) is a hand-hoe used for tillage on a smaller scale while an animal-drawn tiller is used on a larger scale. The load-bearing weight of the Jab planter has been reduced from 9 kg to 6 kg to be lighter and more suitable for female users. The length of the weeder's shallow handle has been increased from 1 m to 1.4 m to reduce stress, especially for women. Hand tillers are used to smash hard pans into the ground.

Soil acidity is a major problem in western Kenya, due to monoculture, especially in the sugarcane region, coupled with the excessive use of chemical fertilizers. The recommended solution is agricultural liming, but this method of application is neither safe nor effective. As a result, the project introduced a more efficient and safer manual lime spreader.

The benefit from using these technologies is that farmers prepare their land in a timely manner as they do not have to depend on cow plow owners who are often overwhelmed with demand for services during peak times. Timely implementation of field activities such as tillage and other field activities helps households mitigate the challenges of climate change such as variations in rainfall. Crop residue and cover crop management has been maximized, resulting in no tillage to conserve soil temperature, water, fertility and leaching capacity, etc.

5. Lessons learnt

5.1 Farmer call centre E-extension model

After promoting mobile phone, video calls, radio voice, TV with a lot of success the project felt the need to have farmer call centers alongside the others to provide extension services where, operators answer farmers' questions at a central location, advice and support smallholder farmers over the phone, using voice and voice call-back to farmers.

5.2 Knowledge management platforms

There were many digital platforms for knowledge management that were operational such as the Ministry websites, County digital platforms, Big Data Systems domiciled at KALRO, Kenya Integrated Agricultural Marketing

Information System (KIAMIS), Livestock Marketing Information System (LMIS), Kenya Agricultural Observatory Platform (KAOP) and Cooperative Management Information System (CMIS).

Sustainability was a key ingredient to long term success in the use of ICTs and appropriate models were put in place to enable the platforms to be used from inception to widespread adoption. There is need for improving access to ICTs particularly internet and computers and creating awareness on use of platforms such as YouTube, Twitter and Farmer Call Centers in accessing agricultural information among farmers.

The availability of CA tools and equipment would accelerate the rate of adoption and sustain their production. Unlike the farmers in the West area, the farmers in the East, Central and Rift Valley areas required more tools. The project trained local artisans to create CA tools in partnership with ATDCs to improve accessibility and affordability. The demand for AC tools and devices has been triggered by increasing awareness. Longer project lead time and improved adoption awareness.

Youth participation in agriculture is minimal and can revolutionize agricultural production as they have greater potential to access and apply modern technologies as they are more open to new ideas, new technologies and practices compared to adult farmers.

Author details

Flora Ajwera[1], George Ayaga[2*], David Kersting[1], Patrick Mudavadi[2], John Manyengo[2], Christopher Mulielie[3], Eboche Dave Khasakhala[4], Pius Cheboi[3] and Daniel Mukhwana[5]

1 GIZ ProSoil, Kisumu, Kenya

2 Kenya Agricultural and Livestock Research Organization, (KALRO) Alupe, Busia, Kenya

3 Ministry of Agriculture (MoA), Kenya

4 Ministry of Agriculture, Siaya, Kenya

5 WHH, Bungoma, Kenya

*Address all correspondence to: ayagag@yahoo.com

References

[1] Dallimer M, Stringer LC, Orchard SE, Osano P, Njoroge G, Wen C, et al. Who uses sustainable land management practices and what are the costs and benefits? Insights from Kenya. Land Degradation and Development. 2018;**29**(9):2822-2835. DOI: 10.1002/ldr.3001

[2] Ziadat FM, Zdruli P, Christiansen S, Caon L, Monem MA, Fetsi T. An overview of land degradation and sustainable land Management in the near East and North Africa. Sustainable Agriculture Research. 2021;**11**(1):11. DOI: 10.5539/sar.v11n1p11

[3] AbdelRahman MA. An overview of land degradation, desertification and sustainable land management using GIS and remote sensing applications. Rendiconti Lincei. 2023;**34**(3):767-808. DOI: 10.1007/s12210-023-01155-3

[4] Timpte M, Montana J, Reuter K, Borie M, Apkes J. Engaging diverse experts in a global environmental assessment: Participation in the first work programme of IPBES and opportunities for improvement. Innovation: The European Journal of Social Science Research. 2018;**31**:S15-S37. DOI: 10.1080/13511610.2017.1383149

[5] Van Beek C, Noij G, Duivenbooden N, Van Heesmans H, Van Til R. More Food from Fertile Grounds Report. Alterra Wageningen (UR); 2013

[6] Tully K, Sullivan C, Weil R, Sanchez P. The state of soil degradation in sub-Saharan Africa: Baselines, trajectories, and solutions. Sustainability (Switzerland). 2015;7(6):6523-6552. DOI: 10.3390/su7066523

[7] Gebreselassie S, Kirui OK, Mirzabaev A. Economics of land degradation and improvement in Ethiopia. In: Economics of Land Degradation and Improvement - A Global Assessment for Sustainable Development. Switzerland: Springer Nature; 2015. DOI: 10.1007/978-3-319-19168-3_14

[8] Bjornlund V, Bjornlund H, Van Rooyen AF. Why agricultural production in sub-Saharan Africa remains low compared to the rest of the world–a historical perspective. International Journal of Water Resources Development. 2020;**36**(suppl. 1):1-34. DOI: 10.1080/07900627.2020.1739512

[9] Shiferaw BA, Okello J, Reddy RV. Adoption and adaptation of natural resource management innovations in smallholder agriculture: Reflections on key lessons and best practices. Environment, Development and Sustainability. 2009;**11**:601-619. DOI: 10.1007/s10668-007-9132-1

[10] Kirui OK, Mirzabaev A, von Braun J. Assessment of land degradation 'on the ground' and from 'above.'. SN Applied Sciences. 2021;**3**(3):1-13. DOI: 10.1007/s42452-021-04314-z

[11] Mwangi V, Owuor S, Kiteme B, Giger M, Jacobi J, Kirui O. Linking household food security and food value chains in north west Mt Kenya. Sustainability (Switzerland). 2020;**12**(12):1-15. DOI: 10.3390/su12124999

[12] Kenya. Population and Housing Census: Volume II i: Vol. II. Nairobi, Kenya: Government Printer; 2019

[13] KNBS. Kenya National Bureau of Statistics. Center for Development and Environment. Nairobi, Kenya: Government Printer; 2016

[14] Kiragu S, Flohr A. Sustainable land Management in Western Kenya – Lessons learnt and future directions. Workshop Report. Potsdam. 2016;**June**:1-16

[15] Mulei JM, Onkware AO, Otieno DF. Vegetation community structure and diversity in swamps undergoing anthropogenic impacts in Uasin Gishu County, Kenya. African Journal of Ecology and Ecosystems. 2016;**3**:175-184

[16] Kogo BK, Kumar L, Koech R. Impact of land use/cover changes on soil erosion in western Kenya. Sustainability (Switzerland). 2020;**12**(22):1-17. DOI: 10.3390/su12229740

[17] Giller KE, Hijbeek R, Andersson JA, Sumberg J. Regenerative Agriculture: An agronomic perspective. Outlook on Agriculture. 2021;**50**(1):13-25. DOI: 10.1177/0030727021998063

[18] Kanyenji GM, Oluoch-Kosura W, Onyango CM, Ng'ang'a SK. Prospects and constraints in smallholder farmers' adoption of multiple soil carbon enhancing practices in Western Kenya. Heliyon. 2020;**6**(3):e03226. DOI: 10.1016/j.heliyon.2020.e03226

[19] Lal R. Soil degradation by erosion. Land Degradation and Development. 2001;**12**(6):519-539. DOI: 10.1002/ldr.472

[20] Sanchez PA, Jama BA. Soil fertility restoration is taking place in Eastern and Southern Africa. In: Integrated Crop Nutrient Management in sub-Saharan Africa. Wallingford, UK: CABI; 2002. pp. 23-45

[21] Dallimer M, Stringer LC, Orchard SE, Osano P, Njoroge G, Wen C. Economics of land degradation (ELD) Kenya study. In: Costs and Benefits of Sustainable Soil Fertility Management in Western Kenya. Report for the Economics of Land Degradation Initiative. 2016. 40 p

[22] Otieno HMO. An assessment of the status of degraded soils and management options in Kenya. International Journal of Innovative Approaches in Agricultural Research. 2021;**5**(2):241-256 DOI: 10.29329/ijiaar.2021.358.9

[23] Keesstra S, Mol G, de Leeuw J, Okx J, Molenaar C, de Cleen M, et al. Soil-related sustainable development goals: Four concepts to make land degradation neutrality and restoration work. Land. 2018;**7**(4):20. DOI: 10.3390/land7040133

[24] Liu T, Bruins RJF, Heberling MT. Factors influencing farmers' adoption of best management practices: A review and synthesis. Sustainability (Switzerland). 2018;**10**(2):1-26. DOI: 10.3390/su10020432

[25] Coenen L, Grillitsch M, Hansen T, Miörner J, Moodysson J. An innovation system framework for system innovation policy: The case of Strategic Innovation Programs (SIPs) in Sweden. In: Papers in Innovation Studies. Lund University, CIRCLE - Centre for Innovation Research; 2017

[26] Eswaran H, Lal R, Reich PF. Land degradation: An overview. Response to land degradation. In: Proceedings of 2nd International Conference on Land Degradation and Desertification. Khonkaen, Thailand; 2019. pp. 20-35

[27] Rai PK, Singh JS. Invasive alien plant species: Their impact on environment, ecosystem services and human health. Ecological Indicators. 2020;**111**:106020

[28] GoK. National Wildlife Conservation and Management Policy. Nairobi: Ministry of Environment and Natural Resources; 2017

[29] Romanelli C, Cooper D, Campbell-Lendrum D, Maiero M, Karesh WB, Hunter D, et al. Connecting Global Priorities: Biodiversity and Human Health, a State of Knowledge Review. World Health Organization and Secretariat for the Convention on Biological Diversity. Geneva, Switzerland: WHO; 2015. p. 360. DOI: 10.13140/RG.2.1.3679.6565

[30] Alila P, Atieno R. Agricultural policy in Kenya: Issues and processes. In: A paper for the Future Agricultures Consortium workshop, Institute of Development Studies. Kenya: Institute for Development Studies, University of Nairobi; 2006

[31] Hanna N. Promising approaches to development challenges. In: Making Development Work. UK: Routledge; 2018. pp. 283-310

[32] Cambodia. Agricultural and rural development. In: Working Paper No. 55. Phnom Penh, Cambondia: CDRI Publication; 2011

[33] Bisht IS, Pandravada SR, Rana JC, Malik SK, Singh A, Singh PB, et al. Subsistence farming, agrobiodiversity, and sustainable agriculture: A case study. Agroecology and Sustainable Food Systems. 2014;**38**(8):890-912. DOI: 10.1080/21683565.2014.901273 ISSN 2168-3565. S2CID 154197444

[34] Ghafoor Awan D, Ghafoor A wan A. Relationship between environment and sustainable economic development: A theoretical approach to environmental problems. International Journal of Asian Social Science. 2013;**3**(3):741-761. Available from: http://www.aessweb.com/journal-detail.php?id=5007

[35] CoG. The Lake Region Economic Blueprint: A Better Life, 2020. Available from: https://cog.go.ke/phocadownload/reports/Lake_Basin_web.pdf

[36] GoK. National Agricultural Soil Management Policy Ministry of Agriculture, Livestock, Fisheries and Co-Operatives. 2021. Available from: https://kilimo.go.ke›uploads›2021/01›Draf

[37] de Brauw A, Bulte E. African farmers, value chains and agricultural development: An economic and institutional perspective. In: Palgrave Studies in Agricultural Economics and Food Policy. London, UK: Palgrave Macmillan; 2021

[38] Webber CM, Labaste P. Building Competitiveness in Africa's Agriculture: A Guide to Value Chain Concepts and Applications. Washington, DC: World Bank Group; 2009

[39] De Janvry A, Sadoulet E. Achieving success in rural development: Toward implementation of an integral approach. Agricultural Economics. 2005;**32**:75-89

[40] Rondot P, Collion MH, editors. Agricultural Producer Organizations: Their Contribution to Rural Capacity Building and Poverty Reduction. Washington, DC: World Bank Group; 2001

[41] Mubichi-Kut F, Quinhentos M, Findeis J. Speaking up: Gender [in] equality in agricultural development. In: Research Brief: USAID and the Feed the Future Soybean Innovation Lab. Division of Applied Social Sciences, University of Missouri-Columbia; 2018

[42] Kenya: The Constitution of Kenya [Kenya]. 2010. Available from: https://www.refworld.org/docid/4c8508822.html

[43] Fairhurst GT, Uhl-Bien M. Organizational discourse analysis (ODA):

Examining leadership as a relational process. The Leadership Quarterly. 2012;**23**(6):1043-1062

[44] Higuchi Y, Yamanaka Y. Knowledge sharing between academic researchers and tourism practitioners: A Japanese study of the practical value of embeddedness, trust and co-creation. Journal of Sustainable Tourism. 2017;**25**(10):1456-1473. DOI: 10.1080/09669582.2017.1288733

[45] Heckert A, Forsythe LP, Carman KL, Frank L, Hemphill R, Elstad EA, et al. Researchers, patients, and other stakeholders' perspectives on challenges to and strategies for engagement. Research Involvement and Engagement. 2020;**6**(1):1-18. DOI: 10.1186/s40900-020-00227-0

Chapter 3

Enhancing Agriculture through Strategic Tillage and Soil Management: Unleashing Potential for Sustainable Farming

Mahima Dixit, Debabrata Ghoshal, Sanjeev Kumar and Debashis Dutta

Abstract

Modern agriculture relies on strategic tillage and soil management to improve soil health, crop yield, and environmental impact. Innovative tillage methods reduce soil disturbance and use alternative soil management methods. Poor tillage reduces soil health and productivity. Some of them include soil erosion, runoff losses, compaction, organic matter deuteration, and nutrient losses. For long-term environmental sustainability, it is important to recognize the risks of improper tillage and implement sustainable soil management methods that reduce soil disturbances, conserve organic matter, improve soil structure, and promote soil stability. In the modern day, mechanization and industrialisation have greatly impacted soil health and ecological balance. This chapter provides a brief review of strategic tillage and soil management concepts, benefits, and challenges. It highlights the shift from traditional tillage to balanced soil management strategies. Strategic tillage improves soil health, conserves resources, decreases erosion, and ensures agricultural resilience and sustainability by limiting soil disturbance, retaining organic matter, and improving fertilizer management. The chapter also highlights aspects cover crops, precision agriculture, and organic farming for soil quality and resource efficiency. This chapter begins to explore the importance and implications of strategic tillage and soil management in modern agriculture.

Keywords: soil management, tillage, sustainability, crop productivity, climate change

1. Introduction

Globally, the agricultural, farming sectors and environmental sustainability are inherently reliant on each other. However, they have undergone significant transformations by mechanization of agriculture, which includes advancements in technology, expansion of operations, socioeconomic progress, alterations in consumption habits, and the looming ignorance of environmental deterioration. About half of Earth's arable land is now used for farming, ranching, or grazing. However, farmers and land managers have altered agricultural ecosystems through resource extraction and wasteful use, in an effort to boost food production for local consumption and export. According to statistics, approximately 2.37 billion individuals, accounting for nearly one-third of the global population, in the 2020 had insufficient access to food due to a lack of resources [1]. Consequently, the challenges of food and nutrition insecurity, in conjunction with climate change, have emerged as the foremost critical and pressing concerns in the pursuit of the Sustainable Development Goals and the goals that are delineated in Agenda 2063.Currently, nations across the globe, both developed and developing, are confronted with pressing and interrelated challenges. These challenges pertain to the critical issues of ensuring food security for a rapidly growing population, as well as the restoration of the environment and preservation of natural resources. In an era marked by population growth, climate change, and escalating environmental concerns, the quest for sustainable agricultural practices has become imperative. At the heart of this endeavor lies scientific tillage, a crucial approach to land cultivation that strives to strike a harmonious balance between agricultural productivity and ecological preservation. Unlike traditional, indiscriminate tillage methods, scientific tillage is an evidence-based, technologically driven approach that recognizes the profound impact of human activities on the planet. By harnessing cutting-edge research and innovative techniques, scientific tillage plays a pivotal role in maintaining sustainability in agriculture, safeguarding soil health, mitigating environmental degradation, and securing the future of food production for generations to come. Consistent use of conventional tillage has been shown to degrade soil structural stability, soil biological characteristics, and nutrient storage and supply, according to growing experimental evidence [2, 3]. After the harvest of Rabi and Kharif season crops, conventional tillage entails a series of mechanical procedures to prepare the soil for the next planting season. These operations include deep plowing, deep disking, ripping, shallow-type workings, and fine seedbed preparation. Subsequently, a designated period of fallow is implemented to facilitate the capture of moisture prior to the commencement of planting the subsequent crop. This method leads to the exposure of a bare soil surface, making it susceptible to erosion caused by wind and water. Additionally, heavy rainfall can result in high compaction, necessitating the need to loosen the soil again for weed control and to enhance moisture absorption during subsequent rain events. Tillage strategies change soil physicochemical characteristics, consequently modifying crop yields [4]. The ongoing degradation of soil in arable land, caused by improper tillage practices, has a notable impact on the physical, chemical, and biological properties of the soil. This leads to a reduction in the storage of soil organic carbon (SOC), as well as a decline in crop productivity and the value of ecosystem services within agricultural systems [5]. There is a prevailing belief that the primary factor contributing to the historical depletion of soil organic carbon (SOC) in North America is soil disturbance caused by tillage. It is widely acknowledged that a significant increase in SOC sequestration can be achieved by transitioning from conventional plowing to less intensive techniques commonly referred to as conservation tillage. Soil tillage modifications bring in a change in soil structure, and therefore, a strategic method of tilling the soil according to the soil type, structure, and environmental condition can prove an effective method to combat carbon losses and other land degradation happening [6]. The soil health plays a vital role in the managing critical zone of the Earth. To effectively work toward the attainment of the United Nations' Sustainable Development Goals (SDGs), it is imperative to uphold fundamental ecosystem services such as food production, plant growth, animal habitat, environmental preservation, carbon sequestration, and overall environmental sustainability. Soil degradation has been observed in various locations worldwide as a result of faulty agricultural practices and lack of knowledge about the new technologies. To effectively accomplish the Sustainable Development Goals (SDGs) within the designated timeframe of 2030, it may be necessary to adopt and implement more sustainable approaches to the utilization and management of soils, surpassing current practices. In this study, we demonstrate the importance of prioritizing the arena of sustainable soil use and management. Specifically, we emphasize the need to focus on the multifunctional tillage practices and their interdisciplinary linkages with major issues concerned with soil health and environment sustainability.

2. Tillage systems

Tillage systems refer to various methods and practices used in agriculture for preparing the soil for planting, managing crop residues, and controlling weeds. These systems involve different degrees of soil disturbance and manipulation to create an optimal environment for crop growth and maximize agricultural productivity. Soil organisms' natural habitats are strongly influenced by the tillage technique used because of the profound effects that system has on the soil's physical and chemical environment. Tillage techniques can affect the soil's moisture, temperature, airflow, and degree of crop residue integration. The changes in soil physical conditions have both direct and indirect impacts on the climate and ecosystem dynamics (**Figure 1**). A significant challenge in the field of soil ecology research involves comprehending the effects of management practices on the intricate interactions among various organisms within the soil community. Soil manipulations alter the living community of soil in several ways. The microorganisms in soil play a crucial role by performing essential functions such as enhancing soil structure, facilitating nutrient cycling, and facilitating degrading of organic matter [8]. The fundamental motivation for reducing tillage intensity and to shift toward innovative methods is to lessen the financial

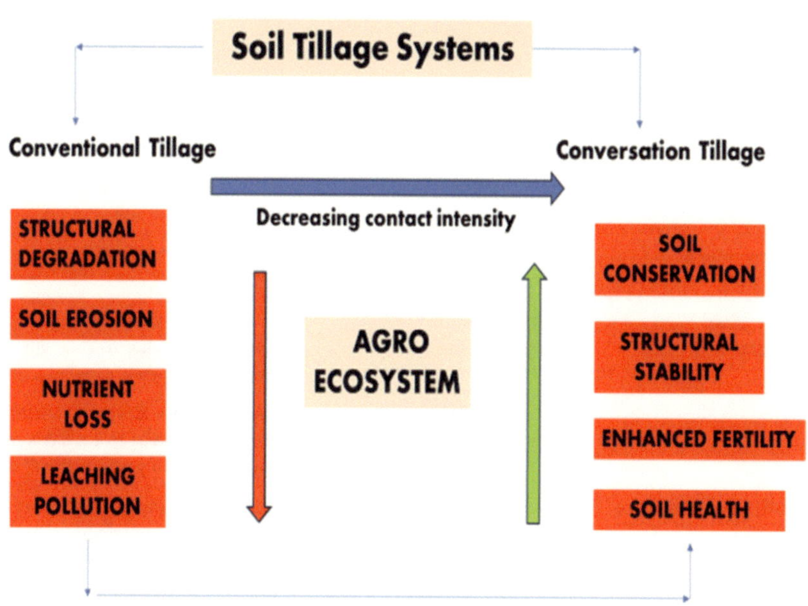

Figure 1.
Tillage systems as an interwoven element in managing the agroecosystems [7].

burden of tillage in the context of commercial agriculture and to reduce the amount of soil degradation sometimes associated with excessive tillage techniques as well as the indigenous techniques of tillage that cause major problems in the soil–environment dynamics as compared to the innovative techniques (**Table 1**). Tillage needs vary moderately according to the crop and soil type. It may not always be necessary or desirable to till an entire field at once. Row crops including corn (*Zea mays* L.), sugar beets (*Beta vulgaris* L.), and oilseed rape (*Brassica napus* L.) can have their tillage restricted to certain areas of the field, a method known as strip or zone tillage. The process of seedbed preparation can be limited to the specific area where the seeds will be planted, while the rest of the land may undergo a reduced or alternative form of tillage. In the non-planted regions, a soil structure with a coarse texture is deliberately maintained to enhance the process of water infiltration. In addition to the aforementioned, crop residues might be left on the soil's surface in the spaces between rows of plants. This method is used to reduce water loss due to evaporation and to protect the soil from wind and precipitation.

The fundamental soil tillage operations encompass either soil inversion or fragmentation. Various tillage implements are employed to carry out tillage operations, operating at specific soil depths. The process of soil inversion is most effectively executed through the utilization of a moldboard plow, whereas soil loosening can be

Aspect	Traditional tillage methods	New innovative tillage techniques
Soil Disturbance	Extensive soil disturbance using plows and heavy equipment	Minimal to no soil disturbance to preserve soil structure and reduce erosion
Crop Residue Management	Crop residues are often buried or removed	Crop residues are left on the surface to improve soil health and moisture retention
Weed Control	Good initial weed control due to soil disruption	May require additional weed management strategies, but can benefit from reduced competition over time
Soil Erosion	Can lead to increased soil erosion	Reduced erosion due to better soil structure and residue cover
Organic Matter Content	May result in a decrease in organic matter over time	Helps build and maintain organic matter content in the soil
Soil Aeration	Provides initial soil aeration	May improve soil aeration over time due to increased soil health
Water Conservation	Less effective in conserving water	Better water conservation due to reduced soil disturbance and residue cover
Energy and Labor Requirements	Can be labor- and energy-intensive	Often reduces the need for intensive tillage equipment, leading to lower energy and labor requirements
Soil Health and Biodiversity	May negatively impact soil health and biodiversity	Supports soil health and promotes biodiversity due to reduced disturbance and increased organic matter
Adaptability and Challenges	Well-established and widely used, but can be environmentally challenging	May require adaptation in weed and pest management, but promotes long-term sustainability

Table 1.
A comparison between the traditional tillage practices and the innovative tillage practices, the challenges, and the constraints.

accomplished by employing various tillage tools. Rotary cultivators are highly suitable for the purpose of soil mixing, frequently employed for the integration of soil fertilizer and organic amendments into the soil. The presence of plant residues has the potential to impede the functionality of certain tillage implements. In agricultural practices, coulters or disks are frequently employed to effectively sever or redirect crop residue from the immediate tillage zone. The occurrence of frequent or intense vehicular movement within the agricultural area can result in the compaction of the subsoil, which refers to the soil located beneath the depth at which tillage activities are performed. Subsoiling refers to the application of mechanical techniques aimed at loosening the compacted subsoil, as outlined in remedial tillage operations. To ensure optimal results, it is recommended that the subsoiling operation be conducted when the soil exhibits a friable consistency, thereby minimizing the risk of smearing. A subsoil that has been mechanically loosened exhibits a diminished strength and is susceptible to significant recompaction with relative ease.

3. Soil and tillage

The composition of soil consists of both living and nonliving components that are arranged in a vertical profile, characterized by horizontal layers or horizons. Soil serves as the habitat for a wide variety of living organisms, encompassing a multitude of microscopic species such as bacteria, fungi, protozoa, nematodes, and microarthropods [9]. Mechanically altering the soil's physical condition in order to improve plant growth for human and animal food production is known as tillage. Plant nutrients and crop additives can be worked into the soil during tillage operations including bed preparation, planting, and postemergence cultivation for weed control. Tillage needs change from crop to crop, depending on things like soil composition and weather. Damages to the environment and agricultural production systems may result if the intensity of tillage is not enough for the local conditions [10]. The fundamental soil tillage operations encompass soil inversion or fragmentation, as outlined in **Table 1**. Various tillage tools are utilized to perform these operations, operating at specific soil depths. The process of soil inversion is most effectively executed through the utilization of a moldboard plow, whereas soil loosening can be accomplished by employing various tillage tools. Rotary cultivators are highly suitable for the purpose of soil mixing, commonly employed to incorporate soil fertilizer and organic amendments into the soil. Plant residues have the potential to impede the functionality of certain tillage implements. Coulter blades or disks are frequently employed to effectively sever or redirect crop residue from the immediate tillage area. Frequent or excessive traffic in the field can result in soil compaction in the subsoil, which refers to the soil located beneath the tillage depth. Subsoiling refers to the application of mechanical techniques to alleviate compaction in the subsoil layer (**Figure 1**, **Table 1**). For optimal subsoiling, it is recommended to conduct the operation when the soil has a water content that allows for friability, thus preventing smearing. A subsoil that has been mechanically loosened exhibits diminished strength and is susceptible to significant recompaction (**Figure 2**).

4. What is strategic tillage?

Strategic tillage is an agricultural practice that involves the targeted and deliberate use of tillage operations in specific areas and at specific times to achieve specific goals.

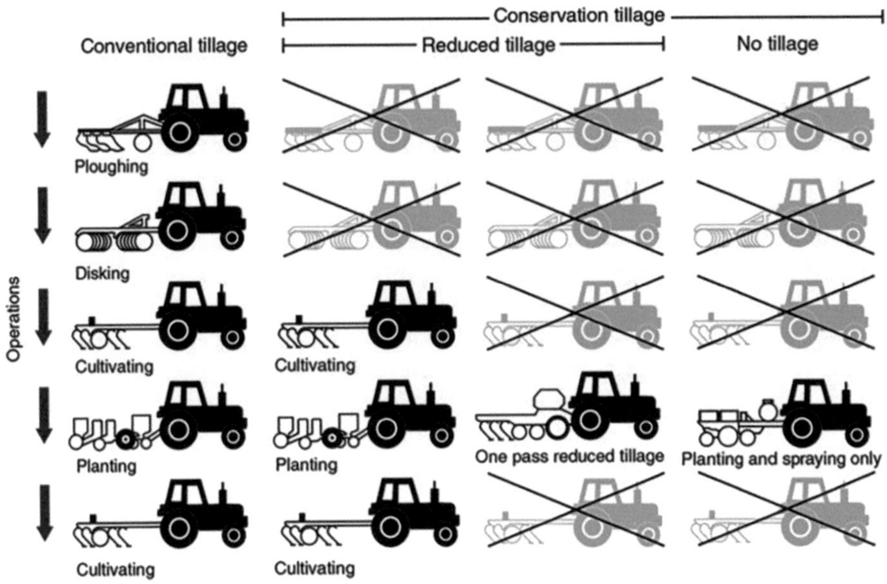

Figure 2.
A pictorial representation of the various tillage operations that provides an overview of the advantages and disadvantages of soil tillage and the various operations in cultivating soil using conventional, reduced and no-till tillage systems, particularly in relation to plowing [11].

Tillage refers to the mechanical manipulation of the soil through plowing, digging, or turning it over. The concept of strategic tillage recognizes that excessive or indiscriminate tillage can have negative effects on soil health, such as soil erosion, loss of organic matter, and disruption of soil structure. Therefore, strategic tillage aims to optimize tillage operations to minimize these negative impacts while maximizing the benefits for crop production [12].

5. Historical background and evolution of strategic tillage practices

The historical development of strategic tillage practices can be categorized into distinct stages. Early agricultural practices involved manual methods such as hand digging, which gradually evolved with the introduction of animal-drawn implements, including the plow. The Industrial Revolution spurred mechanization, leading to increased tillage efficiency but also raising concerns about soil degradation. The emergence of conservation tillage in the mid-20th century marked a significant transition toward sustainable land management, highlighting the need for minimizing soil disturbance and preserving soil health. However, it was the integration of advanced technologies and ecological insights that paved the way for modern strategic tillage practices. The incorporation of precision agriculture tools, remote sensing, and geospatial data analysis enabled farmers to target tillage operations based on field variability, optimizing resource use and minimizing environmental impacts. The adoption of strategic tillage practices aligned with the principles of conservation agriculture, involving reduced tillage, conservation tillage, and minimum tillage and emphasizing soil conservation, nutrient management, and ecosystem resilience.

6. Strategies for implementing strategic tillage for soil management

6.1 Conservation tillage

The phrase "conservation tillage" can be used to describe a wide variety of approaches to soil management. As per the research conducted by Mannering and Fenster in 1983 [13], conservation tillage can be defined as a tillage system that aims to minimize soil or water loss in comparison to conventional tillage practices. This frequently entails the utilization of non-inversion tillage techniques that maintain an adequate layer of residue mulch on the surface of the soil. As per the CTIC, conservation tillage is a designated tillage and planting approach that guarantees a minimum of 30% soil surface coverage through residue after planting. **Table 2** describes the various aspects of conservation tillage. The primary objective of this system is to effectively mitigate water and wind erosion and conservation of soil. In order to keep soil losses to a minimal, it is imperative that certain standards be upheld while taking into account the wind and water conditions. Alternative conservation tillage methods focus on keeping at least 1000 pounds of flat, tiny grain residue equivalent on the soil's surface throughout the critical wind erosion period. Conservation tillage can be broadly mechanized into the following categories that hold a significant difference in soil management.

6.2 Reduced tillage

Targeted and suitable tillage that takes into account the specific needs of the farm is at the center of reduced tillage methods. With reduced tillage, the depth and width of the bed, field, and overall farm are all worked on with a lesser intensity. Due to

No tillage	Ridge tillage	Mulch tillage	Conventional tillage
This method relies on herbicides, cover crops, or both to keep weeds at bay and requires relatively little mechanical seedbed preparation (a narrow band where the seed is sown).	Sweeps, disk openers, coulters, or row cleaners are used to prepare ridges in the seedbed before planting. Between the ridges, the residue is left behind. Herbicides and horticulture are both effective methods for eradicating weeds (CTIC*).	The soil is disturbed prior to planting	Generally, it refers to moldboard plowing (inversion of the soil) followed by a secondary tillage operation such as disking and/or harrowing
With the exception of potential fertilizer injection, no-tillage methods involve not disturbing the soil between harvesting and planting.	From harvest to planting, only fertilizer injections are made to the soil.	Chisels, field cultivators, disks, sweeps, and blades are all examples of tillage implements. Herbicide and/or cultivation are used for weed management.	Both cultivation and the application of herbicides are viable options for weed management.

*The Conservation Technology Information Centre (CTIC).

Table 2.
A comparison between the various tillage systems under conservation tillage.

their beneficial effects on soil and water conservation [14, 15] and their ability to lower the demand for fuel, equipment, and labor [16], the use of reduced tillage practices is on the rise worldwide [17]. Due to the reduced level of soil disturbance associated with reduced tillage as compared to conventional tillage, there is a limited mixing of plant residues, fertilizers, and other soil amendments into the soil. Consequently, plant roots have a tendency to proliferate primarily in the uppermost few centimeters of the soil. In addition, the uppermost layer typically exhibits higher moisture levels, lower temperatures, reduced oxidation, and increased acidity [18–20]. Under these conditions, it is observed that the organic matter content tends to exhibit a slower rate of increase or decrease in comparison to that under conventional tillage practices. In addition, the adoption of straw mulching and reduced tillage practices has a significant influence on the microenvironment of soil microorganisms, consequently affecting the overall sustainability of the environment [21].

6.3 Zero tillage

When referring to a tillage system, the term "zero tillage" is used to indicate that only traffic and seedbed preparation are performed mechanically on the soil. It is the most extreme kind of minimum tillage, a grouping of tillage systems that includes the plow–plant method and other practices that decrease the amount of soil disturbed during cultivation. Compared to conventional tillage (CT), zero tilling (ZT) causes immediate and substantial alterations to the soil pore network. Long periods of time without disturbing the soil allow biotic and abiotic processes to work together to create a stable soil structure [22]. Plant matter is allowed to remain on the soil's surface under a zero-till system. This is crucial in cases where soil erosion prevents farming from being profitable. The observed proportionate increase in the number of medium to tiny pores generated by zero tillage has implications for the soil's water-holding capacity. The act of plowing grassland results in the redistribution of organic matter, whereas zero-tilled sod retains its original deposit of organic matter in close proximity to the soil surface. Baeumer and Bakermans [23] discovered that water content at pF 2 changed more in association with organic matter content than with soil porosity. As a result, the top 6 cm of the zero-tilled soil had a larger water content at pF 2 than that of the plowed soil.

6.4 Minimum tillage

The minimum tillage approach is a soil conservation system that shares similarities with strip-till. Its primary objective is to minimize soil manipulation while still achieving successful crop production. No-till farming is a cultivation technique that avoids soil inversion, as opposed to intensive tillage practices that involve the use of plows to alter the soil structure. In the practice of minimum tillage, primary tillage is entirely omitted, and only a limited amount of secondary tillage is employed. The practice of minimum tillage encompasses various techniques, such as minimizing furrowing, employing organic fertilizers, utilizing biological pest control methods, and reducing reliance on chemical inputs. The application of minimum tillage (MT) in combination with organic farming has proven to be effective in promoting the improvement of the biomass of soil microbe. The microbial community structure experiences a shift toward bacteria within organic farming systems. Bacterial populations demonstrate a more prominent reaction to agricultural practices in comparison to other groups of microorganisms. This response is particularly evident when minimum tillage techniques are employed, as they contribute to the improvement of soil's biological properties. Consequently, these practices indirectly influence the overall health of the soil [24].

7. Conventional vs. conservation: benefits of conservation tillage on soil properties

7.1 Soil physical properties

The impact of conservation tillage on soil physical properties can vary depending on several factors, including soil type, climate, prevalent cropping system, and managerial practices. Nevertheless, conservation tillage, on the whole, facilitates the enhancement of soil integrity, control of erosion, infiltration of water, conservation of moisture and organic matter content, and the reduction of soil compaction, thereby resulting in improved soil physical properties. The effects of conservation tillage on soil properties demonstrate variability, with the choice of a specific system playing a pivotal role in determining these variations. No-till (NT) systems, characterized by the maintenance of extensive surface soil coverage, have been found to induce notable alterations in soil properties, particularly within the upper few centimeters [25]. Numerous studies have indicated that the utilization of NT (no till) has led to a significant enhancement in both saturated and unsaturated hydraulic conductivity. This improvement can be attributed to either the preservation of pore continuity, as observed by Benjamin in 1993 [26], or the facilitation of water flow through a limited number of larger pores. According to Butorac [27], there is evidence indicating that soils with good drainage, a light to medium texture, and low humus content demonstrate an optimal response when exposed to conservation tillage practices, specifically no tillage. According to the research conducted by Lal et al. [28], it has been established that the adoption of NT technologies offers notable advantages in terms of mitigating soil and crop residue disturbance, regulating soil evaporation, and reducing erosion losses. It has been observed that no-till (NT) soils exhibit a greater presence of stable aggregates in the upper soil surface when compared to in tilled soils. As a result, this leads to an increased overall porosity in the NT plots. According to a study conducted by Pagliai et al. [29], minimum tillage was found to enhance the soil pore system when compared to conventional plowing. The observed improvements were characterized by an increase in the storage pores ranging from 0.5 to 50 mm, as well as an increase in the quantity of elongated transmission pores ranging from 50 to 500 mm. The researchers established a correlation between the elevated microporosity observed in minimum tillage soils and an associated rise in soil water content. Consequently, this leads to an increase in enhanced water availability for plant uptake. In a study conducted by McVay et al. [30], it was seen that the topsoil (0–10 cm) under no-till (NT) practices exhibited a greater water-holding capacity or moisture content compared to soil that had undergone plowing. Consequently, in order to enhance soil water retention and optimize water utilization efficiency (WUE), numerous scholars have suggested substituting conventional tillage practices with conservation tillage methods [31, 32].

8. Soil chemical properties

The examination of the enduring effects of conservation tillage on soil chemical properties is a subject of considerable importance to farmers, agronomists, and environmentalists. The implementation of conservation tillage practices leads to the creation of a unique soil environment, which in turn requires modified management approaches when compared to conventional tillage methods. Soil pH, cation exchange capacity (CEC), exchangeable cation levels, and total nitrogen content are among the many of the chemical parameters typically affected by the tillage techniques. Lal [33] found that distinct positive changes in chemical properties of the surface layer tend to occur when employing the no-till method compared to tilled soil. Research by Rasmussen [34] shows that plant residues on the soil's surface contribute to an increase in organic matter inside the topsoil when no-tillage methods are used on a yearly basis. Ismail et al. [35] also documented a notable increase in soil organic carbon (SOC) levels in no-till (NT) soil when compared to in untilled soil. It has been observed that the total nitrogen loss is reduced under the no-till (NT) practice in comparison to under conventional tillage (CT). In tilled lands, the organic carbon and total nitrogen shows a crisp decrease as compared to in land with minimum tillage, which can be attributed to higher rates of mineralization and/or leaching due to the deterioration of soil structure caused by tillage practices; the soil remains unaltered Dalal [36]. The nutrient availability is effectively maintained as the organic matter content is increased in conservation tillage practices. The implementation of conservation tillage practices has been found to have a beneficial effect on the nutrient content of soil. Soils managed with conservation tillage exhibit elevated levels of essential nutrients, including nitrogen (N), phosphorus (P), and potassium (K). The levels of exchangeable calcium (Ca), magnesium (Mg), and potassium (K) in the surface soil are notably elevated under no-till (NT) practices in comparison to under the plowed soil. Conservation tillage systems contribute to the enhancement of nutrient availability for crops by promoting improved soil organic matter content and reducing erosion.

9. Soil biological properties

Conservation tillage practices bring significant changes to soil biological properties compared to conventionally plowed soils. With minimal soil disturbance and increased crop residues left on the soil surface, conservation tillage creates a more favorable environment for soil organisms. One notable impact is the increase in microbial diversity and activity. The continuous supply of organic matter from crop residues fosters microbial growth, leading to enhanced nutrient cycling and improved decomposition of organic materials. This, in turn, boosts nutrient availability for plants, supporting healthier crop growth. Another notable change is the promotion of earthworm and macroinvertebrate activity. The presence of crop residues provides these soil organisms with a constant food source and shelter, leading to higher populations and increased soil bioturbation. The significance of earthworms and macroinvertebrates in the context of soil aeration, organic matter breakdown, and nutrient cycling cannot be overstated. These organisms are instrumental in enhancing soil structure and promoting nutrient availability. The implementation of conservation tillage practices has been found to have a positive impact on the accumulation of soil organic matter. The reduced soil disturbance and slower decomposition rates in conservation tillage systems allow organic matter to build up over time. This accumulation enhances soil biological activity and nutrient retention, promoting a more vibrant and resilient soil ecosystem. Organic carbon is a crucial component that serves as a significant energy source and effectively governs microbial activity within soil [37]. The utilization of conventional tillage techniques is known to have a significant impact on the preservation and maintenance of organic carbon (OC) levels. The primary source of soil enzymes is derived from microorganisms when compared to from plants and animals [38]. The tillage practices chosen directly and indirectly impact the biodynamics of soil. The presence of carbon in CT promotes the acceleration of soil enzyme activities, thereby contributing to the enhancement of soil biological health [39].

Moreover, conservation tillage systems foster a more significant presence of mycorrhizal fungi. The reduced disturbance and continuous presence of crop residues create favorable conditions for mycorrhizal symbiosis with plant roots. Mycorrhizal associations enhance nutrient uptake, particularly phosphorus and micronutrients, further supporting plant growth and health [40]. The utilization of computed tomography (CT) has been observed to yield positive effects and foster harmonious interactions with soil biological components. The minimal alteration of the land plays a crucial role in facilitating the establishment and development of communities, thereby fostering a harmonious equilibrium among various biological components [41, 42]. Overall, conservation tillage practices lead to a more biologically active and diverse soil ecosystem compared to conventionally plowed soils. These changes in soil biological properties contribute to improved soil health, increased nutrient availability, enhanced carbon sequestration, and, ultimately, sustainable agricultural practices.

10. Tillage as a tool for modification of soil ecology and modification in soil environment—a case study

The adoption of effective tillage practices is an essential requirement for the long-term maintenance of soil health and the optimization of crop production [43]. Conservation tillage (no tillage, reduced, and strip) increases soil microbial activity, moisture content of the soil, organic matter, stability of aggregates, cation exchange capacity, and crop production [44, 45]. The impact of tillage practices on soil chemical and physical properties, as well as the quality of fruits and crop yield, as documented by some studies conducted in watermelon and rice-maize cropping systems [46, 47]. By increasing plant water usage efficiency and decreasing irrigation water and labor use compared to traditional farming, conservation tillage, which makes use of permanent beds and strip tillage, has the potential to boost farmers' net income and benefit–cost ratio. In their experiment, Kaisi et al. [48] discovered that tillage intensity greatly decreased soil macro- and microaggregate stability. Soil organic matter rose by 0.17% and soil accessible P by 3.8% in the topsoil (0–20 cm) when conservation tillage procedures were used instead of conventional tillage [49]. Maintaining crop residues on the top soil surface layer (full cover, no till; partial cover, strip tillage) can also reduce soil erosion and increase soil moisture content [50, 51]. Nevertheless, there is ongoing debate regarding the impact of different tillage practices, including conservation and conventional methods, on soil microorganisms and physical properties. The utilization of conservation tillage practices has been observed to result in an increase in the abundance of microorganisms in the soil crop system, though the amount of effectiveness is outlined by several ecological factors and field conditions [52, 53]. A 10-year study on the effects of different tillage methods on tomato yields indicated that conventional tillage (moldboard plow) resulted in 52% fewer nematodes than conservation strip tillage [54]. The fall measurements of nematode population composition in a strip tillage field revealed the following: 1900 bacterivorous, 40 fungivorous, 283 omnivorous, 37 predatory, and 1869 root-feeding (plant parasitic) worms. Moldboard plow soil had a total of 407 bacterivores, 67 omnivores, 14 predators, and 350 root-feeding nematodes, but no fungivores were found. However, the presence of nematodes does not always indicate that the soil is healthy or functioning well. For instance, in comparison to conventional tillage, strip tillage resulted in significant increases of total bacteria (49 percent), active bacteria (27 percent), active and total fungi (37 percent), and total nematodes (275 percent) [55]. In contrast to conventional tillage, strip tillage has been demonstrated in the same study to increase (9-fold) root-feeding nematodes (harmful to plant roots) and lower (possibly) soil nutrient content (P and NO_3^--N) [56]. Nematode feeding behavior and reproductive rate, both of which react rapidly to changes in the rhizosphere, may be a potent reason to this. Nematodes benefit more from conservation tillage because it promotes the growth of more beneficial microorganisms (bacteria and fungi) than conventional methods. Higher weed pressure management is a major drawback of conservation tillage approaches, especially no-tillage systems [57]. Soil bulk density and compaction (increased penetrometer resistance) can also increase in no-till conditions. By contrast, conventional tillage procedures improve soil structure by increasing air and water circulation, reducing compaction, and alleviating weed pressure, while also incorporating crop leftovers and nutrients [58]. Over relatively short periods of time, tillage causes substantial biophysical and biochemical changes. Direct effects on creatures include death, injury, or exposure to predators, in addition to the indirect effects of habitat destruction (**Figure 1**). Alterations in the tillage regime can impact the predominance of certain species, the number of populations, and the diversity of communities in the soil.

11. Future directions and research needs

One potential research gap in precision tillage is the optimization of tillage practices for sustainable agriculture. While precision tillage techniques have gained popularity due to their potential for reducing soil erosion, improving water infiltration, and increasing crop yields, there is still a need for further research to refine and optimize these practices.

Here are a few specific research areas within precision tillage/strategic tillage that could be explored:

1. Depth control: Investigating the optimal tillage depth for different soil types, crop rotations, and field conditions could help minimize soil disturbance while still achieving desired outcomes such as weed control and residue management. Additionally, studying the effects of varying tillage depths on soil compaction and root development could provide valuable insights for precision tillage practices.

2. Variable rate tillage: Developing algorithms and decision support tools for variable rate tillage could enhance precision in the field. This involves identifying and mapping spatial variability in soil properties, such as soil texture, organic matter content, and compaction, and using this information to prescribe varying

tillage depths or intensities across the field. Evaluating the agronomic and economic impacts of variable rate tillage could help optimize its implementation.

3. Conservation tillage and cover cropping integration: Investigating the integration of precision tillage with conservation practices like cover cropping could lead to more sustainable and environmentally friendly systems. Research could focus on determining the most effective timing and methods for terminating cover crops using precision tillage techniques, as well as assessing the impacts on soil health, nutrient cycling, and weed suppression.

4. Soil quality and carbon sequestration: Examining the long-term effects of precision tillage on soil quality and carbon sequestration is essential for assessing its sustainability. Research could investigate the changes in soil organic matter content, microbial activity, soil structure, and nutrient availability under different precision tillage systems. Furthermore, exploring the potential of precision tillage to mitigate greenhouse gas emissions and contribute to climate change adaptation could be a valuable area of study.

5. Economic analysis and machinery development: Assessing the economic feasibility and cost-effectiveness of precision tillage systems, including the required machinery and equipment, is crucial for adoption by farmers. Conducting economic analyses that consider the investment costs, operational expenses, and potential returns associated with precision tillage practices could provide valuable insights for farmers and policymakers. Moreover, exploring opportunities for developing and improving precision tillage machinery, such as sensors, actuators, and automation technologies, could further enhance the efficiency and effectiveness of these practices.

12. Conclusion

Throughout the centuries, tillage has been employed as a method to create a suitable seedbed for cultivating crops, as well as to manage weed growth and integrate various substances such as fertilizers, pesticides, manures, and other amendments. In the early phases, the process of converting native ecosystems through tillage played a crucial role in extracting nutrients from soil organic matter. This extraction process allowed for the utilization of naturally fertile soils to cultivate crops, thus meeting the needs of a burgeoning population with limited resources. Nevertheless, it is important to note that tillage practices have had a detrimental impact on soil organic matter and soil structure. This has resulted in an acceleration of soil erosion and the disruption of the intricate network of soil organisms [59]. In places where soil is intensively and continually farmed, the results of soil deterioration brought on by soil preparation owing to the unchecked exploitation of agricultural systems are plain to see [60].

The process of soil inversion and pulverization through repeated tillage has been found to enhance the decomposition of organic matter, thereby impacting the physical, chemical, and biological properties of the soil. These properties are considered crucial indicators of soil quality [44]. Over the course of the last three decades, there have been significant advancements in land management systems aimed at reducing the necessity for soil tillage. Methods such as no tillage are employed to maintain plant residues on the soil surface, thereby safeguarding the soil against erosion. If crop rotation is implemented in the reduced tillage system, it appears to be economically competitive with conventional tillage. Long-term benefits to local and global ecosystems are greater with the decreased tillage approach. A slower, pulsed release of nutrients for plant uptake results from the shift in decomposition rate from a microbial to a fungal route, and crop residue retention generally has a favorable effect on soil biotic complexity. When compared to regular tillage, crop residue retention reduces CO_2 emissions to the environment and preserves the soil's useful chemical and physical qualities for the long term. High crop productivity in traditional high-input agroecosystems is accomplished not through the breakdown of SOM, but rather by the use of synthetic fertilizers (mostly nitrogen, potassium, and phosphorus). Reduced retention of nitrogen and phosphorus (N and P) fertilizers and large losses of SOM in conventionally tilled systems are major problems [61].

Author details

Mahima Dixit[1,2], Debabrata Ghoshal[3], Sanjeev Kumar[1] and Debashis Dutta[3]*

1 Sardar Vallabhbhai Patel University of Agriculture and Technology, India

2 ICAR-Indian Institute of Farming System Research, Meerut, India

3 ICAR – Indian Agricultural Research Institute, New Delhi, India

*Address all correspondence to: debashisdutta74@gmail.com

References

[1] Mrabet R. Sustainable agriculture for food and nutritional security. In: Farooq M, Gogoi N, Pisante M, editors. Sustainable Agriculture and the Environment. Cambridge, MA, USA: Academic Press; 2023. pp. 25-90

[2] Wang JB, Chen ZH, Chen LJ, Zhu AN, Wu ZJ. Surface soil phosphorus and phosphatase activities affected by tillage and crop residue input amounts. Plant, Soil and Environment. 2011;**57**(6):251-257

[3] Xu J, Han H, Ning T, Li Z, Lal R. Long-term effects of tillage and straw management on soil organic carbon, crop yield, and yield stability in a wheat-maize system. Field Crops Research. 2019;**233**:33-40

[4] Sadiq M, Li G, Rahim N, Tahir MM. Sustainable conservation tillage technique for improving soil health by enhancing soil physicochemical quality indicators under wheat mono-cropping system conditions. Sustainability. 2021;**13**(15):8177

[5] Raimondo M, Nazzaro C, Marotta G, Caracciolo F. Land degradation and climate change: Global impact on wheat yields. Land Degradation & Development. 2021;**32**(1):387-398

[6] Baker JM, Ochsner TE, Venterea RT, Griffis TJ. Tillage and soil carbon sequestration—What do we really know? Agriculture, Ecosystems & Environment. 2007;**118**(1-4):1-5

[7] Tebrugge F, During R-A. Reducing tillage intensity—A review of results from a long-term study in Germany. Soil & Tillage Research. 1999;**53**(1):15-28

[8] Kladivko EJ. Tillage systems and soil ecology. Soil and Tillage Research. 2001;**61**(1-2):61-76

[9] Somasundaram J, Dalal R, Lal R. Conclusions: perspectives on sustainable soil management. In: Sustainable Soil Management: Beyond Food Production. Cambridge Scholars Publishing; 2023. p. 322

[10] Lamandé M, Munkholm LJ, Børresen T. Soil Tillage☆, Reference Module in Earth Systems and Environmental Sciences. Elsevier; 2023

[11] Hallett PD, Bengough AG. Managing the soil physical environment for plants. In: Gregory PJ, Nortcliff S, editors. Soil Conditions and Plant Growth. Chichester, UK: Wiley-Blackwell; 2013. pp. 238-268

[12] Pearsons KA, Omondi EC, Zinati G, Smith A, Rui Y. A tale of two systems: Does reducing tillage affect soil health differently in long-term, side-by-side conventional and organic agricultural systems? Soil and Tillage Research. 2023;**226**:105562

[13] Mannering JV, Fenster CR. What is conservation tillage? Journal of Soil and Water Conservation. 1983;**38**(3):140-143

[14] Holland JM. The environmental consequences of adopting conservation tillage in Europe: Reviewing the evidence. Agriculture, Ecosystems & Environment. 2004;**103**(1):1-25. Available from: https://www.grainsa.co.za/conventional-vs-minimum-tillage-and-its-effect-on-soil

[15] Vogeler I, Horn R, Wetzel H, Krümmelbein J. Tillage effects on soil strength and solute transport. Soil and Tillage Research. 2006;**88**(1-2):193-204

[16] Derpsch R. The extent of conservation agriculture adoption

worldwide: Implications and impact. In: Proceedings of the 3rd World Congress on Conservation Agriculture, Nairobi, Kenya, 3-7 October 2005. Harare: ACT; 2005

[17] El Titi A. Implications of soil tillage for weed communities. In: El Titi A, editor. Soil Tillage in Agroecosystems. Boca Raton, FL: CRC Press; 2003. pp. 147-185

[18] Blevins RL, Thomas GW, Cornelius PL. Influence of no-tillage and nitrogen fertilization on certain soil properties after 5 years of continuous corn 1. Agronomy Journal. 1977;**69**(3):383-386

[19] Doran JW. Soil microbial and biochemical changes associated with reduced tillage. Soil Science Society of America Journal. 1980;**44**(4):765-771

[20] Rice CW, Smith MS, Blevins RL. Soil nitrogen availability after long-term continuous no-tillage and conventional tillage corn production. Soil Science Society of America Journal. 1986;**50**(5):1206-1210

[21] Du C, Li L, Effah Z. Effects of straw mulching and reduced tillage on crop production and environment: A review. Water. 2022;**14**(16):2471

[22] Wardak DLR, Padia FN, de Heer MI, Sturrock CJ, Mooney SJ. Zero tillage has important consequences for soil pore architecture and hydraulic transport: A review. Geoderma. 2022;**422**:115927

[23] Baeumer K, Bakermans WAP. Zero-tillage. Advances in Agronomy. 1974;**25**:77-123

[24] Gupta A, Singh UB, Sahu PK, Paul S, Kumar A, Malviya D, et al. Linking soil microbial diversity to modern agriculture practices: A review. International Journal of Environmental Research and Public Health. 2022;**19**(5):3141

[25] Anikwe MAN, Ubochi JN. Short-term changes in soil properties under tillage systems and their effect on sweet potato (Ipomea batatas L.) growth and yield in an Ultisol in South-Eastern Nigeria. Soil Research. 2007;**45**(5):351-358

[26] Benjamin JG. Tillage effects on near-surface soil hydraulic properties. Soil and Tillage Research. 1993;**26**(4):277-288

[27] Butorac A. Conservation tillage in eastern Europe. In: Conservation Tillage in Temperate Agroecosystems. CRC Press; 2017. pp. 357-374

[28] Lal R, Reicosky DC, Hanson JD. Evolution of the plow over 10,000 years and the rationale for no-till farming. Soil and Tillage Research. 2007;**93**(1):1-12

[29] Pagliai M, Vignozzi N, Pellegrini S. Soil structure and the effect of management practices. Soil and Tillage Research. 2004;**79**(2):131-143

[30] McVay KA, Budde JA, Fabrizzi K, Mikha MM, Rice CW, Schlegel AJ, et al. Management effects on soil physical properties in long-term tillage studies in Kansas. Soil Science Society of America Journal. 2006;**70**(2):434-438

[31] Fabrizzi KP, Garcıa FO, Costa JL, Picone LI. Soil water dynamics, physical properties and corn and wheat responses to minimum and no-tillage systems in the southern pampas of Argentina. Soil and Tillage Research. 2005;**81**(1):57-69

[32] Silburn DM, Freebairn DM, Rattray DJ. Tillage and the environment in sub-tropical Australia—Tradeoffs and challenges. Soil and Tillage Research. 2007;**97**(2):306-317

[33] Lal R. Long-term tillage and maize monoculture effects on a tropical Alfisol in western Nigeria. I. Crop yield and soil physical properties. Soil and Tillage Research. 1997;**42**(3):145-160

[34] Rasmussen KJ. Impact of ploughless soil tillage on yield and soil quality: A scandinavian review. Soil and Tillage Research. 1999;**53**(1):3-14

[35] Ismail I, Blevins RL, Frye WW. Long-term no-tillage effects on soil properties and continuous corn yields. Soil Science Society of America Journal. 1994;**58**(1):193-198

[36] Dalal RC. Long term trends in total nitrogen of a vertisol subjected to zero-tillage, nitrogen application and stubble retention. Soil Research. 1992;**30**(2):223-231

[37] Kotroczó Z, Juhos K, Biró B, Kocsis T, Pabar SA, Varga C, et al. Effect of detritus manipulation on different organic matter decompositions in temperate deciduous forest soils. Forests. 2020;**11**(6):675

[38] Bandick AK, Dick RP. Field management effects on soil enzyme activities. Soil Biology and Biochemistry. 1999;**31**(11):1471-1479

[39] Nugroho PA, Juhos K, Prettl N, Madarász B, Kotroczó Z. Long-term conservation tillage results in a more balanced soil microbiological activity and higher nutrient supply capacity. International Soil and Water Conservation Research. 2023;**11**:528-537

[40] Cookson WR, Murphy DV, Roper MM. Characterizing the relationships between soil organic matter components and microbial function and composition along a tillage disturbance gradient. Soil Biology and Biochemistry. 2008;**40**(3):763-777

[41] Engell I, Linsler D, Sandor M, Joergensen RG, Meinen C, Potthoff M. The effects of conservation tillage on chemical and microbial soil parameters at four sites across Europe. Plants. 2022;**11**(13):1747

[42] Kurm V, Schilder MT, Haagsma WK, Bloem J, Scholten OE, Postma J. Reduced tillage increases soil biological properties but not suppressiveness against Rhizoctonia solani and Streptomyces scabies. Applied Soil Ecology. 2023;**181**:104646

[43] Jabro JD, Stevens WB, Evans RG, Iversen WM. Tillage effects on physical properties in two soils of the Northern Great Plains. Applied Engineering in Agriculture. 2009;**25**:377-382

[44] Cannell R, Hawes J. Trends in tillage practices in relation to sustainable crop production with special reference to temperate climates. Field Crops Research. 1994;**30**:245-282

[45] Gozubuyuk Z, Sahin U, Ozturk I, Celik A, Adiguzel MC. Tillage effects on certain physical and hydraulic properties of a loamy soil under a crop rotation in a semi-arid region with a cool climate. Catena. 2014;**118**:195-205

[46] Gathala MK, Timsina J, Islam MS, Rahman MM, Hossain MI, Harun-Ar-Rashid M, et al. Conservation agriculture-based tillage and crop establishment options can maintain farmers' yields and increase profits in South Asia's rice–maize systems: Evidence from Bangladesh. Field Crops Research. 2015;**172**:85-98

[47] Sahu P, Singh D, Prabha R, Meena K, Abhilash P. Connecting microbial capabilities with the soil and plant health: Options for agricultural sustainability. Ecological Indicators. 2019;**105**:601-612

[48] Al-Kaisi M, Douelle A, Kwaw-Mensah D. Soil microaggregate and macroaggregate decay over time and soil carbon change as influenced by different tillage systems. Journal of Soil and Water Conservation. 2014;**69**(6):574-580

[49] Shao Y, Xie Y, Wang C, Yue J, Yao Y, Liu W. Effects of different soil conservation tillage approaches on soil nutrients, water use and wheat-maize yield in rainfed dry-land regions of North China. European Journal of Agronomy. 2016;**81**:37-45

[50] Celik A, Altikat S, Way T. Strip tillage width effects on sunflower seed emergence and yield. Soil and Tillage Research. 2013;**131**:20-27

[51] Mullins GL, Alley SE, Reeves DW. Tropical maize response to nitrogen and starter fertilizer under strip and conventional tillage systems in Southern Alabama. Soil and Tillage Research. 1998;**45**:1-15

[52] Leskovar D, Othman Y, Dong X. Strip tillage improves soil biological activity, fruit yield and sugar content of triploid watermelon. Soil and Tillage Research. 2016;**163**:266-273

[53] Yang S, Kim M, Seo Y, Choi K, Lee S, Kwak Y, et al. Soil microbial community analysis of between no-till and tillage in a controlled horticultural field. World Journal of Microbiology and Biotechnology. 2012;**28**:1797-1801

[54] Overstreet L, Hoyt G, Imbriani J. Comparing nematode and earthworm communities under combinations of conventional and conservation vegetable production practices. Soil and Tillage Research. 2010;**110**:42-50

[55] Meena VS, Verma JP, Verma SK, Sihag SK. Towards the prime response of manure to enhance nutrient use efficiency and soil sustainability a current need: A book review. Journal of Cleaner Production. 2016;**112**:1258-1260

[56] Dotaniya M, Meena V, Basak B, Meena R. Potassium uptake by crops as well as microorganisms. In: Meena VS, Maurya BR, Verma JP, Meena RS, editors. Potassium Solubilizing Microorganisms for Sustainable Agriculture. New Delhi, India: Springer; 2016. pp. 267-280

[57] Sharratt B, Zhang M, Sparrow S. Twenty years of conservation tillage research in subarctic Alaska II. Impact on soil hydraulic properties. Soil and Tillage Research. 2006;**91**:82-88

[58] Zhang YJ, Wang S, Wang H, Ning F, Zhang Y, Dong Z, et al. The effects of rotating conservation tillage with conventional tillage on soil properties and grain yields in winter wheat-spring maize rotations. Agricultural and Forest Meteorology. 2018;**263**:107-117

[59] Papendick RI, Parr JF. No-till farming: The way of the future for a sustainable dryland agriculture. Annals of Arid Zone. 1997;**36**:193-208

[60] Derpsch R, Moriya K. Implications of on-tillage versus soil preparation on sustainability of agricultural production. Advances in Geoecology. 1998;**31**:1179-1186

[61] Lal R. Soil Conservation, Reference Module in Life Sciences. Elsevier; 2017

Chapter 4

Strategic Tillage for Improved Soil Health and Nutrient Dynamics

Omkar Singh, Uday Pratap Shahi, Debashish Dutta, Shivangi, Vishnu D. Rajput and Abhishek Singh

Abstract

Excessive conventional tillage can degrade important aspects of soil quality and health over time. Strategic tillage offers a focused solution to address priority soil limitations with minimal disturbance. This chapter reviews the current science on how strategic tillage affects key soil physical, chemical, and biological factors relevant to soil function and crop yields. In contrast, strategic tillage involves the targeted, occasional, and shallow use of tillage equipment to address specific observed soil constraints without general field disruption. Potential advantages of strategic tillage discussed include improved soil structure, increased infiltration and rooting depth, better incorporation of amendments, disruption of compaction, enhanced soil biological activity and carbon storage, increased nutrient availability, and improved crop yields. To minimize disturbance and maintain soil function, strategic tillage should be integrated with other conservation practices like cover crops and crop rotation. Criteria for selecting appropriate tillage equipment and practices based on crop, soil, and climate factors are explored. Ongoing site-specific evaluation and refinement of strategic tillage practices using crop yields and soil analysis is emphasized. Despite needing more research, strategic tillage shows promise as a precise soil management approach that maximizes productivity and resilience while balancing periodic focused tillage with principles of soil health.

Keywords: cover crops, crop productivity, crop rotation, tillage, soil health, soil quality

1. Introduction

Tillage is one of the most fundamental practices in agriculture, used to prepare soil for planting, control weeds, manage crop residues, and incorporate amendments. However, excessive and inappropriate tillage can lead to a range of issues that degrade soil health over time. Conventional intensive tillage using implements like moldboard plows and heavy disk harrows can cause serious soil degradation through compaction, erosion, loss of soil organic matter, disruption of soil structure, and depletion of soil nutrients [1, 2]. This decline in soil health under conventional tillage regimes has raised concerns about the long-term sustainability and productivity of farmland under crop production.

In response, there has been a strong push in recent decades to reduce overall tillage intensity and adopt conservation tillage systems that retain protective crop residue cover. However, it is increasingly recognized that eliminating tillage entirely may not be optimal or practical in all cropping systems and soil conditions. Some degree of strategic tillage may provide benefits by disrupting compacted layers, incorporating amendments, managing heavy residue loads, and creating seedbeds for adequate plant stands [3, 4]. The goal of strategic tillage is to balance the need for tillage with the need to protect soil health in a planned, precision approach.

Strategic tillage can be defined as the targeted, occasional use of tillage implements in specifically defined locations, depths, and times to address priority limiting factors while minimizing disturbance of the soil [4, 5]. Unlike routine conventional tillage over entire fields, strategic tillage is applied only where evidence indicates it is necessary. For example, deep tillage may be used to shatter compacted subsoil layers that limit root growth and water infiltration, while shallow zone tillage could incorporate surface-applied fertilizers or break up crusted soils that are prone to poor seedling emergence. The type, timing, frequency, and intensity of strategic tillage varies based on the cropping system context, soil conditions, climate patterns, and specific production goals. Overall, the mindset shifts from a "more tillage is better" approach to using minimal, strategic tillage only as needed to accomplish the intended soil improvements and agronomic benefits.

Multiple studies have now documented that strategic tillage systems can provide equivalent or greater crop yields compared to conventionally tilled soils, while building soil health and retaining many of the environmental advantages of no-till systems [3, 4, 6]. Importantly, the benefits of strategic tillage may be most pronounced in certain soil types and cropping systems. For example, heavy clay soils that are prone to compaction may benefit from occasional deep loosening to improve internal soil drainage and root penetration [7]. Strategic shallow tillage can also help mitigate some challenges faced in long-term no-till systems, such as nutrient stratification, cool wet seedbeds, and heavy residue loads. However, improper implementation of strategic tillage, such as too frequent or aggressive use, could certainly negate benefits and accelerate soil degradation. Overall, evidence to date suggests strategic tillage may offer a useful integration of conventional and no-till approaches to balance multiple soil and agronomic needs, but site-specific conditions must guide appropriate application.

One major goal and purported benefit of strategic tillage is improving overall soil health, which encompasses the chemical, physical, and biological properties that drive soil function and productivity [8]. Soil health degradation associated with intensive tillage is extensive, including loss of soil organic matter, disruption of soil structure, reduced water infiltration, increased compaction, greater erosion, and decreases in biological diversity and activity in the soil food web [1, 9, 10]. Strategic tillage aims to minimize these detrimental effects while still utilizing tillage where evidence shows that it can rectify specific limitations. For example, occasional deep tillage in compacted subsoil zones may improve soil structure and water movement, increase rooting depth, stimulate microbial activity, and improve soil carbon storage while only impacting a small portion of the soil profile [3, 11]. Integrating strategic tillage with practices like cover crops, rotations, and reduced surface disturbance may support many components of soil health while targeting the most critical soil limitations [12].

Soil nutrient dynamics are a key aspect of soil health that may be impacted by strategic tillage implementations. Nutrients like nitrogen and phosphorus are essential for crop productivity and need to be carefully managed for availability, efficiency, and prevention of losses through erosion, leaching, or gas emissions. Conventional intensive tillage often results in depletion of native soil nutrients and disruption of biological nutrient cycling, necessitating ever-greater inputs of synthetic fertilizers to maintain yields [1]. Strategic tillage may influence multiple aspects of soil nutrient status and behavior relative to no-till systems, including soil organic matter mineralization, soil nutrient stratification, gaseous emissions, and interactions with applied nutrient sources [3, 13]. For example, occasional shallow incorporation of surface residues and amendments may increase nutrient contact with actively cycling soil biology to enhance plant availability and reduce buildup of nutrients like phosphorus at the surface [1]. But, excess disturbance could also accelerate soil organic matter losses and nitrogen mineralization [11].

2. Definition of soil health and its key components

Soil health is a fundamental concept in agriculture and environmental science, representing the overall well-being and quality of a soil ecosystem. It refers to the capacity of soil to perform its functions effectively and sustainably over time. Soil health is a holistic concept that encompasses various physical, chemical, and biological properties of soil. Its key components include:

1. *Soil Structure:* Soil structure refers to the arrangement and organization of soil particles into aggregates or clumps. A healthy soil typically has well-developed granular or crumb structure. Good soil structure provides adequate pore spaces for air and water movement, root penetration, and the habitat for soil organisms.

2. *Organic Matter:* Organic matter in soil consists of decomposed plant and animal materials, such as humus. It plays a crucial role in soil health by improving soil structure, water-holding capacity, and nutrient retention. Organic matter serves as a source of energy and nutrients for soil microbes and enhances soil fertility.

3. *Microbial Activity:* Soil is teeming with microorganisms, including bacteria, fungi, protozoa, and nematodes. These microorganisms are essential for soil health as they are involved in nutrient cycling, organic matter decomposition, and the suppression of soil-borne diseases. A diverse and active microbial community indicates a healthy soil ecosystem.

3. The relationship between soil structure, organic matter, and microbial activity

The health of these key components of soil is interrelated, creating a dynamic ecosystem within the soil:

- *Soil Structure and Organic Matter:* Good soil structure is often associated with higher levels of organic matter. Organic matter acts as a binding agent, promoting the formation of stable soil aggregates. These aggregates, in turn, protect organic matter from rapid decomposition, contributing to soil stability and improved structure.

- *Microbial Activity and Organic Matter:* Microbes thrive on organic matter, breaking it down into simpler compounds. This process, known as decomposition, releases nutrients in forms that plants can absorb. In return, plants provide organic matter to the soil through root exudates, supporting microbial populations.

- *Microbial Activity and Soil Structure:* Soil microbes contribute to soil aggregation by producing sticky substances that bind soil particles together. They also create tunnels and channels through their movements, improving soil aeration and water infiltration. Active microbial communities help maintain soil structure.

4. Importance of soil health for crop productivity and environmental sustainability

Soil health is pivotal for agricultural sustainability and ecosystem stability:

- *Crop Productivity:* Healthy soils promote vigorous root growth, efficient nutrient uptake, and optimal water retention. This results in increased crop yields and improved crop quality. Farmers with healthier soils are often more resilient to droughts and other weather extremes.

- *Environmental Sustainability:* Soil health is closely tied to environmental sustainability. Healthy soils reduce the risk of soil erosion, nutrient runoff, and water pollution. They also sequester carbon, mitigating climate change. Furthermore, diverse and active soil microbial communities play a role in preventing diseases and enhancing pest control in agricultural systems.

5. Impacts of conventional tillage on soil health

Since the origins of agriculture, tillage has been used to prepare seedbeds, control weeds, and incorporate amendments. However, conventional intensive tillage using frequent disturbance through implements like moldboard plowing and heavy disking can seriously impair soil health over time. Tillage breaks up soil structure, disrupts biological communities, and exposes soil organic matter to decomposition, resulting in a cascade of physical, chemical, and biological degradations.

5.1 Physical impacts

One of the most prominent physical impacts of intensive tillage is increased soil compaction and degradation of soil structure [2, 14]. Frequent tractor passes and inversion by moldboard plowing form plow pans and destroys soil aggregation, reducing macroporosity for drainage, aeration, and root growth [7, 15]. With poor structure, tillage also leaves soil much more vulnerable to compaction and wheel traffic effects [16]. Loss of stable aggregation from organic binding agents and disruption of soil biology impair soils' ability to regain structure [10].

In addition, conventional tillage leaves soil nearly bare between crops, dramatically increasing susceptibility to erosion from wind and rainfall. Intensive tillage pulverizes surface aggregates and residues that protect soil from raindrop impact and overland flow [17]. Erosion removes fertile topsoil and organic matter, reduces infiltration, and causes further structural decline. Tillage also speeds evaporation of soil moisture, requiring more frequent passes for seedbed preparation that perpetuates compaction issues.

Formation of surface crusts is another common physical impact in conventionally tilled soils. Destruction of surface aggregates combined with loss of protection from residues and minimal biological activity near the surface often causes crusting and poor seedling emergence [7]. Hard crusts reduce infiltration and increase runoff and erosion. Emerging seedlings can be cut off or plants may expend crucial energy breaking through crusted layers.

Overall, intensive conventional tillage degrades the fundamental physical properties needed for soil functionality. The resulting problems with compaction, structure loss, erosion risks, and crusting diminish root growth, drainage, and plant establishment which are critical for agricultural productivity and soil sustainability.

5.2 Chemical impacts

Tillage also substantially alters key chemical properties and processes in soil systems. Most notably, conventional intensive tillage generally causes significant declines in soil organic matter and overall carbon stocks [3, 11]. Frequent inversion and disturbance expose previously protected organic matter to microbial decomposition and release CO_2 into the atmosphere [17]. Reduced carbon inputs from less residue return also deplete organic matter over time. Lower organic matter negatively impacts structure, reduces nutrient and water-holding capacities, and decreases soil biology.

Nutrient balances and availability dynamics are also disrupted by conventional tillage regimes. Native soil nutrient reserves become depleted over years of crop removal without adequate replenishment from organic cycling or additions of fertilizers and amendments [1]. Nutrients like nitrogen and phosphorus can be rapidly lost through leaching and gas emissions when bound up in soil organic matter that is decomposed through tillage exposure. Soluble nutrients are also readily lost when tillage causes erosion. More intensive fertilizer inputs are thus needed to maintain crop nutrition in degraded tilled soils. However, fertilizer use efficiency is often impaired by poor soil structure and biological activity as well.

Improper pH levels resulting from conventional tillage may further impact soil chemical processes and crop growth. Acidification is common, stemming largely from greater nitrate leaching when ammonium fertilizers are rapidly converted to nitrate, and reduced liming due to minimal surface residue incorporation [18]. Lower pH substantially influences nutrient availability, solubility of heavy metals and aluminum, and microbial communities (**Table 1**).

5.3 Biological impacts

Soil biology represents a foundational component of overall soil health that is significantly altered by frequent conventional tillage [19]. The extensive physical disturbance and changing soil environmental conditions under intensive tillage regimes reduce soil biological diversity, activity, and community resilience [20]. Fungal communities are particularly affected due to their sensitivity to disturbance, while bacteria increase in dominance [21].

Impact category	Description	Consequences
Physical impacts		
Soil compaction	Excessive machinery and tillage equipment weight can compact soil, reducing pore spaces and aeration.	• Impedes root growth and water infiltration. • Decreases soil oxygen levels, impacting beneficial soil organisms.
Soil erosion	Soil is vulnerable to erosion due to exposed soil surfaces and disrupted soil structure.	• Loss of topsoil, which contains valuable nutrients. • Sediment runoff into water bodies, harming aquatic ecosystems.
Soil crusting	Aggressive tillage can break down soil aggregates, leading to the formation of a hard crust on the soil surface.	• Impedes seedling emergence and root penetration. • Reduces water infiltration, increasing surface runoff.
Chemical impacts		
Organic matter decline	Frequent and intensive tillage accelerates the decomposition of organic matter in the soil.	• Diminished soil organic carbon levels. • Reduced soil structure and water-holding capacity.
Nutrient losses	Disturbance of the soil can result in the leaching of essential nutrients such as nitrogen and phosphorus.	• Nutrient imbalances in the soil. • Environmental pollution through nutrient runoff.
Biological impacts		
Reduction in microbial diversity	Disruption of the soil habitat affects the diversity and abundance of soil microorganisms.	• Loss of beneficial soil microbes. • Reduced nutrient cycling and organic matter decomposition.
Reduction in microbial activity	Soil disturbance can decrease microbial activity, as microorganisms are sensitive to disturbance.	• Slower decomposition of organic matter. • Limited availability of nutrients for plants.

Table 1.
Comprehensive analysis of the impacts of conventional tillage on soil health.

Biomass, abundance, and diversity of soil fauna including protozoa, nematodes, earthworms, and arthropods generally decline with tillage intensity, which removes surface residues and kills organisms directly through mechanical disruption [22, 23]. Declines in mycorrhizal associations further impact plant and root ecology. Reduced soil macrofauna and fungal activity negatively affect soil structure development and stabilization. Lower organic matter under tillage also cuts off the energy source driving soil food web integrity and function.

Overall biological activity including nutrient mineralization and immobilization, carbon transformations, and pest suppression suffers under intensive tillage regimes [24]. For example, earthworms and arbuscular mycorrhizal fungi play key roles in soil carbon storage; their reduction with tillage decreases soil carbon sequestration [10, 25]. Impaired biology diminishes critical services like nutrient cycling, water regulation, and disease suppression.

In essence, conventional tillage degrades soil biological function at many levels, from microbial communities and soil fauna diversity to critical services that support agricultural productivity and ecological health. Regenerating soil biology requires a systemic approach including reduced tillage, increased plant diversity, and enhanced organic matter inputs.

6. Principles and benefits of strategic tillage

Strategic tillage represents a paradigm shift from routine, intensive conventional tillage to a targeted, minimalist approach focused on addressing priority limitations. The core principles and purported benefits of strategic tillage include:

6.1 Goals

The goals of strategic tillage align with improving overall soil function to support crop productivity. Key aims are enhancing soil structure, increasing water infiltration and rooting depth, managing heavy residue loads, incorporating amendments, and disrupting compaction or plow pans [2, 3]. Unique from blanket conventional tillage, the intent is pinpointed remediation of soil conditions limiting productivity in a particular field.

6.2 Timing

Strategic tillage involves careful timing of tillage based on in-field conditions, crop needs, and climate factors. For example, deep tillage may be most effective when soils are dry enough for shattering compaction but moist enough for aggregation reformation [26]. Strategic tillage is also matched to crop growth stages, like disrupting soil crusts prior to planting or clearing heavy residues after harvest. Overall, timing focuses on achieving the intended soil improvements when conditions allow for effective, minimally disruptive tillage.

6.3 Types

A range of tillage types and implements can strategically address different soil limitations. Zone tillage like strip tillage or vertical tillage disturbs narrow slots for planting rather than full inversion [27]. This helps increase biological activity, residue incorporation, and rooting depth in the planting zone while minimizing surface disturbance. Subsoiling, ripping, or other deep tillage tools fracture and mix dense, restrictive subsurface layers 12–18 inches down while leaving surface soil intact [28]. Light incorporation implements like disks, sweeps, and shallow cultivators mix amendments and clear residues from the seed zone while retaining protective surface cover between rows [17].

6.4 Benefits

When applied judiciously, targeted strategic tillage can impart multiple benefits to soil health and crop productivity compared to intensive conventional tillage. These include:

- *Increased Soil Organic Matter and Carbon Sequestration:* Reduced disturbance increases organic matter inputs while occasional mixing can stimulate microbial activity and aggregation to protect carbon [3, 4].

- *Improved Soil Biology:* Strategic tillage in rotations or during cover crop windows causes less frequent disruption, increasing soil fauna and microbial diversity more than annual tillage [24].

- *Reduced Erosion:* Retaining surface cover and only disturbing portions of the soil protects against erosion while creating conditions for better infiltration and soil structure to resist erosive forces [17].

- *Better Nutrient Cycling:* Strategic mixing and root zone incorporation can place nutrients in biologically active zones without excessive mineralization of organic matter [1].

- *Increased Productivity:* Timely loosening of root-limiting compacted layers, residue clearing, and zone building increases root growth, drainage, and plant establishment (**Table 2**) [26].

Benefit of strategic tillage	**Description**	**Consequences**
Improved soil structure	Strategic tillage practices minimize soil disturbance, preserving soil aggregates and organic matter.	• Enhanced water infiltration and root penetration. • Increased soil aeration and porosity.
Enhanced water infiltration	Reduced soil compaction and surface disturbance lead to improved water infiltration rates.	• Better water retention, reducing the risk of drought stress. • Minimized surface runoff and erosion.
Reduced erosion and runoff	Strategic tillage helps maintain soil cover and structure, reducing erosion and sediment runoff.	• Preservation of topsoil and valuable nutrients. • Minimization of sediment pollution in water bodies.
Increased soil organic matter	Minimal soil disturbance allows organic matter to accumulate, enriching the soil with nutrients and carbon.	• Enhanced soil fertility and nutrient availability. • Improved microbial activity and nutrient cycling.
Enhanced microbial diversity	Reduced disruption of soil habitats promotes a diverse microbial community.	• Improved nutrient cycling and organic matter decomposition. • Enhanced plant-microbe interactions for nutrient uptake.
Weed suppression	Strategic tillage can disrupt weed growth cycles, reducing weed competition with crops.	• Decreased weed pressure and the need for herbicides. • Improved crop yield and quality.
Energy and cost savings	Less intensive tillage reduces fuel and machinery costs.	• Increased economic viability for farmers. • Reduced greenhouse gas emissions.
Environmental sustainability	Strategic tillage practices align with sustainable agriculture principles, minimizing negative environmental impacts.	• Preservation of biodiversity and ecosystem services. • Reduced agricultural contributions to climate change and pollution.

Table 2.
Comprehensive benefits of strategic tillage practices in agriculture.

7. Strategic tillage effects on soil nutrients

In addition to overall soil physical and biological properties, strategic tillage can influence key chemical processes in soils related to plant nutrition. These include impacts on nutrient stratification, availability, fertilizer efficiency, and nutrient cycling.

8. Impacts on nutrient stratification and availability

One common effect of long-term no-till systems is increased stratification of nutrients like phosphorus and potassium at the soil surface, since inputs are not mixed through tillage [1]. While this can facilitate early crop access, it can also lead to localized depletion over time. Occasional strategic tillage may modify these stratified layers and move surface nutrients into the active rooting zone, increasing availability during later growth stages [29]. For example, Abdollahi and Munkholm [3] found that periodic deep loosening of no-tilled soils redistributed potassium down the profile. Similarly, Reicosky and Archer [30] showed that tillage incorporation of composted swine manure reduced surface phosphorus accumulation compared to no-till.

However, excessive disturbance through aggressive strategic tillage could quickly deplete surface nutrient reserves. More moderate vertical tillage may provide a balance, with Schomberg et al. [31] finding that strip tillage increased phosphorus deeper in the root zone compared to no-till, likely improving access during grain filling. In general, occasional strategic tillage shows potential to modify stratified nutrient layers favorably but should be constrained to prevent over-mixing.

Stratified nutrient layers can also influence overall mineralization and nutrient availability for crop uptake. For instance, Franzluebbers [32] found that tillage incorporation of cover crop residues, which contained high N and P concentrations, nearly doubled plant availability of these nutrients compared to surface-retained residues. This illustrates how strategic tillage augmenting buried organic matter decomposition could supplement crop nutrition when coordinated with key growth stages.

9. Interactions with fertilizers and residues

Strategic tillage approaches generally increase fertilizer use efficiency compared to no-till, through better incorporation with soil [1]. For example, Reider et al. [33] found that deep banding of phosphorus fertilizer under strip tillage increased nutrient uptake and yields compared to broadcast fertilizer under no-till, likely due to placement in moist subsurface zones with higher biological activity. However, excessive disturbance could expose nutrients to leaching or gaseous losses before crop utilization.

Occasional tillage can also encourage decomposition of surface residues and release of nutrients like nitrogen for crop use. Wright et al. [34] showed that strip tillage increased nitrogen mineralization from previous cover crop residues compared to no-till, synchronized with peak corn demand. Strategic tillage must balance mineralization and nutrient uncovering with preventing excess loss through disturbance.

10. Maximizing nutrient cycling

By fostering a more active and diverse soil biological community while limiting soil degradation, strategic tillage can enhance internal nutrient cycling [24]. Periodic tillage aerates soils, mixes crop residues, and incorporates amendments to stimulate microbial activity and soil fauna critical to decomposition and nutrient release [3]. For example, Plaza et al. [35] found that occasional chisel plowing increased soil nematode populations and nitrogen mineralization compared to untilled soils under conservation agriculture.

Strategic tillage also encourages development of soil structure and permeability to water flows that facilitate cycling and prevent nutrient losses [26]. Combined with practices like cover cropping, manuring, and reduced chemical inputs, strategic tillage could make soils more self-sufficient through enhanced biological nutrient transformations [1]. However, over-intensive disturbance risks accelerating organic matter and nutrient depletion.

10.1 Implementing strategic tillage in farming systems

While the concepts behind strategic tillage are straightforward, real-world implementation requires integrating it effectively within diverse crop rotations, soil types, and climates. Key considerations include combining strategic tillage with other conservation practices, selecting suitable equipment, and evaluating impacts through soil and crop monitoring.

10.2 Integration with other conservation practices

Rather than a stand-alone approach, strategic tillage should be viewed as one potential tool within a diversified soil health management system [17]. Combining occasional strategic tillage with practices like cover cropping, diverse rotations, and reduced surface disturbance is key to realizing benefits without compromising long-term productivity and resilience [3].

For example, including deep-rooted cover crop mixes can help fracture compaction and improve infiltration between cash crops. This could reduce the intensity and frequency of subsoiling or deep ripping required for the same purposes [7]. Leaving corn or bean residues intact with zone or strip tillage maintains surface cover for erosion control. Rotational diversification disrupts pest and disease cycles improved by occasional full-width tillage. Overall, the goal is to address all aspects of soil function and health using strategic tillage only where other practices are unable to resolve priority limitations.

10.3 Appropriate equipment and techniques

Selecting suitable tillage implements is crucial for focused soil disruption and residue management rather than broad inversion or disturbance [17]. Equipment like strip tillage tools, shallow disk cultivators, specialty plows, and deep rippers allow targeted intervention. Matching implement capabilities like tillage depth, width, and inversion intensity to specific limitations enables sufficient yet minimal soil disturbance.

Proper setup and operation of equipment is also key, for example, using guidance systems for precise subsurface tillage at optimal speeds and depths. Successfully avoiding compaction from excess equipment passes requires caution. Integrating equipment use into controlled traffic patterns can help restrict soil compaction effects [26]. Overall, the mindset must focus on disturbing only the minimal soil volume needed to address verified limitations.

10.4 Evaluation and monitoring

Ongoing assessment through soil testing and crop yield monitoring provides critical feedback for improving strategic tillage implementation over time. Baseline soil samples prior to adopting strategic tillage establish reference conditions for physical, chemical, and biological indicators of soil function [3]. Periodic retesting then helps gauge impacts on organic matter, compaction, fertility, biology, etc. Post-harvest crop yields correlated to strategic tillage actions provide additional real-world measures of agronomic value from addressing soil constraints.

For example, Raper et al. [28] combined deep tillage and in-row subsoiling based on soil resistance profiling and crop yields showing compaction-limited root growth and water infiltration. Zone sampling over years tracked soil carbon changes showing no significant difference between strategic tillage and continuous no-till [4]. This empirical evidence helps refine decisions on which limitations to address, proper tillage types, depths, timing, and frequency for local conditions. In summary, strategic tillage is not a universal prescription but rather one possible tool integrated through systematic soil health management. Combining it with practices enriching soil biology, structure, and nutrient cycling maximizes the potential for strategic tillage to address priority limitations without degrading soils through excessive disturbance. Matching equipment capabilities and operation to pinpointed soil needs minimizes disturbance. Ongoing monitoring via soil testing and crop yield response provides real-world evidence to refine site-specific implementation over time. When integrated using a goal-oriented systems approach, strategic tillage offers opportunities to balance soil function improvements with practical farming needs.

11. Conclusions and future research directions

The concept of strategic tillage represents a promising bridge between intensive conventional tillage and continuous no-till systems, aiming to provide targeted soil improvements while protecting long-term productivity and resilience. However, realizing the potential benefits requires further research and adaptive management.

11.1 Summary of potential benefits

When judiciously implemented, a growing body of evidence suggests that strategic tillage can provide multiple agronomic and environmental advantages compared to both intensive tillage and exclusively no-till systems. Documented benefits include reduced compaction, increased infiltration and rooting depth, incorporation of amendments, enhanced soil biological activity, increased carbon sequestration, improved nutrient availability and cycling, increased fertilizer efficiency, and higher crop yields where soil constraints are alleviated [1, 3, 4]. By remediating key limitations at critical times using minimal soil disturbance, strategic tillage offers opportunities to improve soil function and productivity without the extensive degradation associated with routine conventional tillage.

11.2 Need for continued research

However, there remain many open questions regarding long-term impacts, proper implementation, and applicability across diverse soil types, climates, and cropping systems. Most studies on strategic tillage have evaluated relatively short-term effects over several years. More extended field research is needed to determine if periodic strategic tillage can sustain long-term soil carbon levels, biological activity, and stable aggregation compared to continuous no-till [29]. Quantifying and isolating the effects of strategic tillage frequency, depth, timing, and equipment settings requires further work. There is also a need to identify relevant indicators for when strategic tillage is warranted to address soil constraints versus conditions where no-till management alone is sufficient.

11.3 Outlook for wider adoption

Incorporating strategic tillage on a wider scale requires an adaptive management approach built on the foundation of soil health principles and site-specific conditions [17]. This entails comprehensive evaluation of soils, crops, and climate interactions to determine where strategic tillage aligned with other practices like cover cropping and crop rotation diversification can alleviate priority limitations. Given the complexities of soil systems, growers may be reluctant to depart from either intensive tillage or strict no-till doctrine without strong evidence for strategic tillage benefits in their contexts. Continued on-farm research and demonstration will be critical for wider adoption, along with technical guidance on monitoring soil function and indicators to profitably integrate strategic tillage. With increased knowledge and adaptive expertise, strategic tillage adoption could expand to sustainably enhance agricultural productivity and soil resilience.

Author details

Omkar Singh[1*], Uday Pratap Shahi[1], Debashish Dutta[2], Shivangi[1], Vishnu D. Rajput[3] and Abhishek Singh[4]

1 Sardar Vallabhbhai Patel University of Agriculture and Technology, Uttar Pradesh, India

2 ICAR – Indian Agricultural Research Institute, New Delhi, India

3 Academy of Biology and Biotechnology, Southern Federal University, Russia

4 Yerevan State University, Yerevan, Armenia

*Address all correspondence to: singhomkar.agri@gmail.com

References

[1] Crittenden SJ, Poot N, Heinen M, van Balen DJ, Pulleman MM. Soil physical quality in contrasting tillage systems in organic and conventional farming. Soil and Tillage Research. 2015;**154**:136-144

[2] Degos A, Lin HS, Nguyen ML, Chang AC. Long-term conventional and no-tillage impacts on selected soil physical properties. Soil and Tillage Research. 2020;**197**:104534

[3] Abdollahi L, Munkholm LJ. Tillage system and cover crop effects on soil quality: I. Chemical, mechanical, and biological properties. Soil Science Society of America Journal. 2014;**78**(1):262-270

[4] Veenstra JJ, Horwath WR, Mitchell JP. Tillage and cover cropping effects on aggregate-protected carbon in cotton and tomato. Soil Science Society of America Journal. 2007;**71**(2):362-371

[5] Baker JM, Ochsner TE, Venterea RT, Griffis TJ. Tillage and soil carbon sequestration—What do we really know? Agriculture, Ecosystems & Environment. 2007;**118**(1-4):1-5

[6] Hermans SM, Prins WH, Kok K, van Wart J. Towards understanding relationships between soil and crop yield–A comparison between strategies for soil–crop monitoring. European Journal of Agronomy. 2020;**118**:126074

[7] Chen G, Weil RR. Root growth and yield of maize as affected by soil compaction and cover crops. Soil and Tillage Research. 2011;**117**:17-27

[8] NRCS. Soil Health. 2023. Available from: https://www.nrcs.usda.gov/wps/portal/nrcs/main/soils/health/

[9] Hobbs PR. Conservation agriculture: What is it and why is it important for future sustainable food production? Journal of Agricultural Science. 2007;**145**(2):127

[10] Six J, Elliott ET, Paustian K. Soil macroaggregate turnover and microaggregate formation: A mechanism for C sequestration under no-tillage agriculture. Soil Biology and Biochemistry. 2000;**32**(14):2099-2103

[11] Alvarez R, Steinbach HS. A review of the effects of tillage systems on some soil physical properties, water content, nitrate availability and crops yield in the Argentine Pampas. Soil and Tillage Research. 2009;**104**(1-2):1-15

[12] Peigné J, Ball BC, Roger-Estrade J, David C. Is conservation tillage suitable for organic farming? A review. Soil Use and Management. 2007;**23**(2):129-144

[13] Venterea RT, Baker JM, Dolan MS, Spokas KA. Carbon and nitrogen storage are greater under biennial tillage in a Minnesota corn–soybean rotation. Soil Science Society of America Journal. 2006;**70**(5):1752-1762

[14] Hamza MA, Anderson WK. Soil compaction in cropping systems: A review of the nature, causes and possible solutions. Soil and Tillage Research. 2005;**82**(2):121-145

[15] Logsdon SD, Allmaras RR, Nelson WW, Copeland PJ. Spatial variability of field-measured soil properties relative to site-specific management. Transactions of the ASAE. 1993;**36**(2):417-0425

[16] Soane BD, Ball BC, Arvidsson J, Basch G, Moreno F, Roger-Estrade J. No-till in northern, western and South-Western Europe: A review of problems

and opportunities for crop production and the environment. Soil and Tillage Research. 2012;**118**:66-87

[17] Reicosky DC. Conservation tillage is not conservation agriculture. Journal of Soil and Water Conservation. 2015;**70**(5):103A-108A

[18] Blevins RL, Thomas GW, Cornelius PL. Influence of no-tillage and nitrogen fertilization on certain soil properties after 5 years of continuous corn. Agronomy Journal. 1977;**69**(3):383-386

[19] Kibblewhite MG, Ritz K, Swift MJ. Soil health in agricultural systems. Philosophical Transactions of the Royal Society B: Biological Sciences. 2008;**363**(1492):685-701

[20] Helgason BL, Walley FL, Germida JJ. No-till soil management increases microbial biomass and alters community profiles in soil aggregates. Applied Soil Ecology. 2010;**46**(3):390-397

[21] Six J, Bossuyt H, Degryze S, Denef K. A history of research on the link between (micro) aggregates, soil biota, and soil organic matter dynamics. Soil and Tillage Research. 2004;**79**(1):7-31

[22] Roger-Estrade J, Anger C, Bertrand M, Richard G. Tillage and soil ecology: Partners for sustainable agriculture. Soil and Tillage Research. 2010;**111**(1):33-40

[23] Wang Q, Wang S, Huang Y. Long-term effect of no-tillage on soil organic carbon fractions in wheat-maize cropping system of the North China Plain. Geoderma. 2012;**179**:115-120

[24] Stinner BR, House GJ. Arthropods and other invertebrates in conservation-tillage agriculture. Annual Review of Entomology. 1990;**35**(1):299-318

[25] Kladivko EJ. Tillage systems and soil ecology. Soil and Tillage Research. 2001;**61**(1-2):61-76

[26] Chen G, Drury CF, Reynolds WD. Impacts of 35 years of conservation and conventional tillage management on soil hydraulic properties. Soil and Tillage Research. 2021;**208**:104960

[27] Licht MA, Al-Kaisi M. Strip-tillage effect on seedbed soil temperature and other soil physical properties. Soil and Tillage Research. 2005;**80**(1-2):233-249

[28] Raper RL, Reeves DW, Burmester CH, Schwab EB. Tillage depth, tillage timing, and cover crop effects on cotton yield, soil strength, and tillage energy requirements. Applied Engineering in Agriculture. 2000;**16**(4):379

[29] Venterea RT, Maharjan B, Dolan MS. Fertilizer source and tillage effects on yield-scaled nitrous oxide emissions in a corn cropping system. Journal of Environmental Quality. 2011;**40**(5):1521-1531

[30] Reicosky DC, Archer DW. Moldboard plow tillage depth and short-term carbon dioxide release. Soil and Tillage Research. 2007;**94**(1):109-121

[31] Schomberg HH, Endale DM, Calegari A, Peixoto R, Miyazawa M, Amado TJC. Influence of cover crops and tillage on phosphorus stratification in soils under long-term no-till subsystem. Soil and Tillage Research. 2014;**140**:60-68

[32] Franzluebbers AJ. Water infiltration and soil structure related to organic matter and its stratification with depth. Soil and Tillage Research. 2002;**66**(2):197-205

[33] Reider C, Herdina H, et al. Precision phosphorus management: Do variable

rate application and inclusion of a starter signal increase wheat yield? Agronomy Journal. 2016;**108**(6):2287-2296

[34] Wright SF, Starr JL, Paltineanu IC. Changes in aggregate stability and concentration of glomalin during tillage management transition. Soil Science Society of America Journal. 1999;**63**(6):1825-1829

[35] Plaza C, Courtier-Murias D, Fernández JM, Polo A, Simpson AJ. Physical, chemical, and biochemical mechanisms of soil organic matter stabilization under conservation tillage systems: A central role for microbes and microbial by-products in C sequestration. Soil Biology and Biochemistry. 2013;**57**:124-134

Chapter 5

Effective Microorganisms (EM): A Potential Pathway for Enhancing Soil Quality and Agricultural Sustainability in Africa

Lindani Ncube

Abstract

There has been a rise in campaigns promoting the protection of the environment and the development of green economies. A green economy aims to improve human well-being and social equity while reducing environmental risks and scarcities. It focuses on low carbon emissions, efficient resource usage, and inclusivity. The concept of green economies emerged due to the negative impact of traditional economies on the environment and natural resources. African economies, which largely rely on agriculture, often resort to high-yield crops and chemical fertilizers to meet food demand. However, these practices are costly for smallholder farmers and contribute to environmental degradation. To address this, alternative systems, such as Effective Microorganisms (EM), are being explored. EM are beneficial microorganisms that can enhance crop growth and soil quality. They consist of lactic acid bacteria, photosynthetic bacteria, yeasts, and actinomycetes. Commercially produced EM have shown promising results in enhancing crop growth. However, further research is needed to fully understand their benefits. EM offer a potential solution for organic soil management without harming the environment and are relatively affordable compared to intensive agriculture practices. Therefore, this chapter offers an in-depth focus on EM and their benefits, and potential role in agricultural food production and prospects.

Keywords: effective microorganisms, soil management, mineral fertilizers, organic manure, sustainable agriculture

1. Introduction

Some of the strategies that can be employed to intensify agricultural output include planting high-yielding varieties, the use of synthetic crop protection agricultural chemicals, and the application of synthetic mineral fertilizers. For example, about 50% of increased agricultural production has been heavily linked to the use of synthetic mineral fertilizers [1]. This approach is, however, increasingly becoming unsustainable as it causes soil degradation, and the cost of required inputs is often beyond the financial reach of communal farmers. These communal farmers constitute more than 80% of food producers in developing nations [2].

Alternative systems that are more suitable to sub-tropical and tropical smallholder farmers are the target. EM are thus an alternative for crop growth enhancement. EM are cultured, naturally occurring specially selected microorganisms that are beneficial to soil health with the potential to significantly improve plant growth and soil quality [3]. These organisms are made up of selected species, including predominant populations of lactic acid bacteria, yeasts, a small number of photosynthetic bacteria, and actinomycetes, among others. These microorganisms need to be mutually compatible with one another to coexist in a liquid culture.

Prof Teruo Higa from the University of the Ryukyus, Okinawa, Japan, was the first individual to develop the concept of EM in 1971. The improvement of crop quality, growth, yield, soil quality, and health after the inoculation of plant ecosystem or the soil with EM has been widely reported [4]. The application of the microorganisms has been shown to increase sweet potato, peas, and onion yield by 23, 29, and 31%, respectively [3]. Local microbial isolates are being used to produce different brands of EM in over 40 countries around the world. Several South African countries are now producing and marketing EM products including EMROSA (Pty) Ltd., EMLife, and Efficient Microbes among others. Although several companies now market EM products, their use is not yet widespread, but over the years, some commercial farmers have adopted them with positive results. As such, there is an increase in EM use. However, there is a lack of rigorous research on the usefulness of EM in crop production in South Africa, and Africa in general.

2. General overview of agricultural intensification

As a result of an increase in global population, food and nutrition insecurity is expected to increase especially in developing countries due to a decline in arable land per capita [5]. However, the production of food for human consumption, animal feed, and industrial purposes has increased in developed countries mainly because of an increase in crop yield and cropping intensity rather than because of an expansion of arable land [6].

2.1 Intensification in agriculture and sustainability

The rapid growth of the global population, currently at 8.045 billion, necessitates fiber and food production to be increased without increasing the land for agricultural activities [7]. The phenomenal growth in the global production of food since the mid-1960s has not seen a matching significant improvement in food and nutrition security. The recent Covid-19 pandemic has made things worse. For example, world hunger increased in 2020 due to the COVID-19 pandemic, while undernourishment increased by 1.5% in 2020, with 218 and 418 million of these cases residing in Africa and Asia, respectively [8]. More than a third of the world's population did not have access to adequate food in 2020; 12% were severely food-insecure [9]. The uneven spread of economic wealth between the developing and developed nations of the world arising from modern production systems, which consequently lead to the overlooking of social equity and environmental sustainability, is one of the contributing factors to the aforementioned food insecurity [10]. Increased food production is not sustainable because of environmental, socioeconomic, and financial inefficiencies in the world

[10]. But currently, statistics indicate that sugarcane, wheat, rice, and maize account for nearly half of the global crop production, and this amounted to 9.4 billion tons in 2019 [11]. In 2021, tomatoes (189.13 MMT), onions (106.59 MMT), cucumbers (93.53 MMT), cabbages and other brassicas (71.71 MMT), and eggplants (58.65 MMT) were the top five most produced crops in the world [12]. In general, China, the United States, Brazil, India, Russia, France, Mexico, Japan, Germany, and Turkey are the top ten agriculture-producing countries in the world. Be that as it may be, one school of thought proposes that the issue of food and nutrition insecurity is heavily linked to food losses along the value chain [13]. According to these authors, about a third of the world's food is lost and therefore does not reach the consumer's table.

2.2 Agricultural production in Africa

Incessant droughts emanating from the *El Nino* weather phenomenon, pests and diseases, and low soil nutritional contents, among others, are leading to low crop yields in Africa [1]. But, the use of agrochemicals, high-yielding cultivars, and mineral fertilizers is part of a wider scale of intensive agricultural production system programs that are being implemented to increase production and yield [14]. However, this attempted intensification of agriculture is leading to a massive decline in soil health and fertility, among others, more so in Africa [15]. The main causes of poor soil fertility in Africa include leaching, improper application of mineral fertilizers, soil erosion, low organic manure amendments, and the removal of crop residues, among others [16]. Africa constitutes 60% of the world's arable land and drives most economies by employing most people, consequently contributing about 14% of sub-Saharan Africa's GDP [17]. Africa therefore poses great potential in terms of crop production and yield in the face of droughts. Although the overuse of both organic and inorganic fertilizers has been linked to environmental degradation among other factors, the underapplication of these, harsh climatic conditions, as well as poor soil management and agricultural practices have led to soil degradation in most parts of SSA. This leads to a decline in yields and consequently food and nutrition insecurity in the region. However, Africa produces corn, wheat, and rice in this order, while corn has the widest use and is grown in all ecological zones [18]. The highest yields of corn have been reported in Egypt, Reunion, and Mauritius, while millet and sorghum are cultivated in the dry regions of the Savanna. The cultivation and consumption of rice are on the increase owing to rapid urbanization. In addition, fodder crops are mainly restricted to East and Southern Africa's highland zones as well as North Africa's subtropical zones [18].

2.3 Constraints of intensifying agriculture in Africa

The use of mineral fertilizers has been adopted at a global scale because they significantly increase yields; however, in Africa, their use is limited because of high costs [19]. Most farmers are subsistence-oriented and constitute about 80% of food producers in the region [2]. Consequently, these farmers are unable to purchase highly priced mineral fertilizers. Research has also shown that 64–70% of SSA farmers are unable to adopt alternative soil management techniques because of financial challenges [20].

2.4 Sustainable agricultural production

One of the most critical aspects of alternative soil management practices and studies is sustainability. Sustainability is seen as the ability of farming systems to persist into the future [21]. According to the previous authors, this means that sustainable agriculture is the ability to ensure that farming systems continuously adapt to new and improved strategies while enabling future food demands to be met for posterity. Social, economic, and environmental aspects are some of the many multidimensional characteristics of sustainable agriculture [22]. Because of its positive environmental, economic, and social roles, organic farming is being promoted and consequently expanding rapidly around the world [23]. Soil microbial pools are the pillar of agricultural production in organic systems, more so in nutrient supply [24]. Thus, organic farming has been on the increase over the last few decades at about ca. 20% annually and covering over 24 million hectares around the world [25, 26]. Consequently, organic farming has been adopted as the mainstream practice for some crops [23].

Organic farming lessens environmental stress [27], improves soil biodiversity [28], and enhances the formation of soil structure while improving food safety and quality [29]. The use of biological and organic inputs for pest and disease control as well as nutrient supply are the main drivers of organic farming [30]. Sustainable organic farming depends largely on the build-up of soil microbial pools that serve as a temporary nutrient sink and are accountable for releasing nutrients from organic matter for plant use [23]. Reports indicate that microbial biomass nitrogen contributes to the primary nitrogen source of potentially mineralizable soil nitrogen [31]. Thus, EM ensures maximum returns of soil microbial pools.

3. Effective microorganisms

According to Safwat and Matta [3], EM are essentially a combination of microorganisms that are beneficial to the soil for health. Predominant species of LAB, a small population of photosynthetic bacteria, yeasts and actinomycetes, among others, constitute EM. To co-exist in an aqueous culture, these microorganisms need to be mutually companionable. Some EM include the following:

3.1 Lactic acid bacteria

Lactic acid is produced by LAB from sugars and carbohydrates that are developed by some yeasts and photosynthetic bacteria. Some of the foods that have been produced for decades using LAB include pickles and yoghurt. Because lactic acid is a strong sterilizing compound, it can enhance organic matter decomposition and suppress harmful microorganisms. Lignin and cellulose are some of the materials whose decomposition is promoted by LAB, followed by fermentation, and this process consequently removes undesirable effects of non-decomposed organic matter [32].

3.2 Yeasts

Plant roots, organic matter, and photosynthetic bacteria secrete sugars and amino acids, which are further synthesized into antimicrobials and some beneficial substances that are required for plant growth. Root and active cell division are thus promoted by hormones and enzymes, which are some bioactive compounds, that are produced by the yeasts. These secretions act as useful substrates for effective microbes such as actinomycetes and LAB [33].

3.3 Photosynthetic bacteria

Phototropic or photosynthetic bacteria are a group of self-supporting and independent microorganisms. By using soil heat and sunlight as a source of energy, these bacteria synthesize beneficial substances from harmful gases such as hydrogen sulphide and/or organic matter and root secretions. Plant development and growth-promoting substances such as nucleic acids, amino acids, sugars, and bioactive substances are all developed by these microbes. According to Prescott et al. [33], the metabolites developed by these microbes are absorbed directly into plants and act as substrates for increasing useful populations.

3.4 Actinomycetes

Actinomycetes are microbes that constitute EM, and they disturb the life cycle of insects by reducing their reproduction rate. Metamorphosis is hindered when actinomycetes feed on larvae-produced chitin before the larvae becoming pupae. Actinomycetes are gram-positive and aerobic, hence can be cultivated easily on simple growing media [33].

3.5 Brands of EM

The increasing global popularity of the use of EM as crop production broad-based organic materials is one of agriculture's innovative solutions to farming. Various EM brands have been shown to improve crop quality and growth [34]. Various EM brands are discussed in detail in the succeeding sections.

3.5.1 Multiplied (M-) EM

Stock EM is multiplied to reduce cost. Multiplied-EM is a mixture of stock EM with molasses and water in a ratio of 1:5:94. After mixing the ingredients, the resultant solution is stored in an airtight container for between 10 and 14 days between 20 and 25°C until the pH is 3.7. During this time, the microbes enter a growth phase and multiply to reach a microbial population and composition like that of the stock EM. To maintain the quality of multiplied EM, they should be stored between 10 and 20°C [35, 36].

3.5.2 EM-fermented plant extracts (EM-F.P.E)

Ingredients used for EM-fermented plant extract production: chopped fresh weeds (20L—not compressed); chlorine-free water (16 L); molasses 480 ml (3%); multiplied-EM 480 ml (3%) [36]. Weeds or plants that possess repelling properties, for example, herbs, grass, syringa, clover, and khakibos are used to produce EM – EM-fermented plant extract. Plants are chopped fresh into 2–5 cm pieces and deposited into a sealable bucket during brewing. The multiplied-EM is then mixed with water and the resultant solution is poured into the bucket containing the chopped plant material, and the bucket is closed. After 2–5 days, the fermentation of the plant constituents begins, and a lot of carbon dioxide is generated and released through an air trap. The process goes on until the pH drops below 3.7 at which point the solution is filtered to remove the plant materials [3, 35, 36]. EM-F.P.E. acts as a disease suppressor and repellent, and it also contains vitamins, enzymes, hormones, and amino acids.

The EM-F.P.E. is either sprayed on the crop with a sticker or applied to the soil through irrigation during the growing season [3, 35, 36].

3.5.3 EM 3-in-1

EM 3-in-1 is one of the strongest EM-based insect repellents. This is produced in a similar way as EM-F.P.E. except for the ingredients used. About 400 g each of chili pepper, ginger, and fresh garlic are chopped, and 200 g of black pepper is powdered, while 600 ml of Multiplied-EM and 18 L of water are used [35, 36].

3.5.4 EM-5

EM-5 is a mixture of EM1 (multiplied EM), water, vinegar, strong distillation alcohol (>30%), and molasses. Following ingredient mixing, the resultant is fermented in a tightly closed container for at least 30 days or until carbon dioxide production ceases. During the fermentation process, herbs such as red pepper and garlic are added. Thus, to strengthen the natural immune system against diseases and attacks by destructive pests, the resultant solution is applied to the targeted crop [35, 36].

3.5.5 EM-bokashi

EM-bokash is a mixture of multiplied-EM with quality and fresh organic materials such as wheat bran, fish meal, or rice bran. Following the mixing of the ingredients, the resulting mixture is kept for up to 2 weeks in tightly closed containers.

The final product is used for the following:

- Anerobic decomposition and accelerating fermentation of organic waste materials when making compost.
- To incorporate in animal feed to improve general health and natural immunity [3, 35, 36].

3.5.6 EM-fermented fish (EM-F.F)

To allow smooth plant growth over time and from one season to the next, EM-F.F is added to the soil, thus releasing nutrients such as phosphorus and nitrogen slowly over time. The fish are ground to fine powder during preparation to enable accessibility to microorganisms. Before spraying on the plants, the resultant mixture is combined with multiplied EM [3, 35, 36].

3.5.7 EM-fermented chicken manure

Because EM-fermented chicken manure provides phosphorus and nitrogen to plants, it plays a similar role to that of EM-fermented fish. The ingredients used include an equivalent of 1% bokash, extended EM, and chicken manure.

3.5.8 EM-fermented kitchen garbage

For EM-fermented kitchen garbage, organic waste that is generated in the kitchen is fermented using multiplied EM to produce a nutrient-rich fertilizer for plants. The production procedure for this is like that of EM-F.P.E. production [35, 36].

3.5.9 EM-X

This distinct version of EM liquid has been certified for human intake. This beverage EM (C [EM-X]) is an antioxidant cocktail that is derived from papaya, seaweed, and unpolished rice fermentation with grouped yeast, photosynthetic bacteria, and LAB EM. This is composed of EM and mixed plant extracts. EM-X is constituted of at least 40 minerals and antioxidants (including lycopene, ubiquinone, quercetin, kaempferol, pannexin, ascorbic acid, oryzanol, flavonoids and tocopherol) as well as some bioactive substances including amino acids (such as nicotinamide adenine dinucleotide, I-alanine, l-glutamine, and nicotinamide mononucleotide), peptides, and nucleotides [37]. The consumption of EM-X daily overtime leads to a reduction in free radicals in the body and can potentially lead to the reduction of cancers and boosting of the immune system [35, 36].

3.6 Application of EM

According to Olle and Williams [38], EM are effective for crop production and are environmentally safe. According to previous authors, EM are produced in over 40 countries of the world using local microbial isolates, environmentally friendly, and effective for crop production. Their applications are numerous and diverse, ranging from aquaculture, animal, and crop production as well as environmental management. Various brands of EM are thus applied under different conditions in many ways as discussed in the following sections.

3.6.1 Inoculation of EM into the soil

EM can either be sprayed on plants or be applied as a soil drench during crop production. A dilution ratio of 1:500 multiplied by - EM in water or EM-FKG (kitchen garbage) when applied on the soil is used. For EM-F.C.M or EM-F.F, a dilution ratio of 1:300 is recommended. An equivalent of 2.5 tons of bokash or less is applied to soil per hectare. Applications above 2.5 t ha^{-1} can be detrimental/toxic to the plants because of organic acids, which can potentially damage plant roots. EM-bokash is usually applied 10–14 days before planting and is placed between 10 and 15 cm away from roots [3, 35, 36]. EM in combination with other substances such as molasses has been shown to improve in onion, peas, and sweetcorn [39]. The study of zu Schweinsberg-Mickan and Müller [40] showed that the addition of EM to the soil significantly increased soil microbial activity. A global metadata analysis has previously shown that soil inoculation with EM results in about a 16% yield increase in all crops, but legumes show a superior response [41]. For example, EM inoculation in the soil significantly increased yield, seed quality traits, and leaf photosynthesis in the common bean [42]. In general, a host of studies have shown a significant increase in crop yield after the inoculation of the soil with EM [43–45].

3.6.2 Spraying EM on leaves

The spraying of effective microorganisms on plant leaves serves as a prophylactic spray mainly for insect and disease control. This is done earlier in the season and terminated when the crop has been harvested. A dilution ratio of 1:1000 of multiplied-EM, EM-F.P.E. or EM-5, or a mixture of different EM derivatives, is recommended, although a stronger dilution can also be used [3, 35, 36].

3.6.3 Soaking seeds in EM

Seeds are soaked in 0.1% EM water before planting. Smaller species are soaked for 30 minutes, while the bigger species are soaked for up to 6 hours. Seeds are then dried under the shade to reduce the chances of sticking together [3, 35, 36] and planted in the field.

3.6.4 EM irrigation (fertigation)

Irrigation water is often used to supply multiplied EM or EM derivatives to the soil. A dilution ratio ranging between 1:1000 and 5000 multiplied-EM or EM-F.P.E to water is used [3, 35, 36].

3.6.5 Insect control

Controlling and suppressing pests through the introduction of beneficial EM to the crop's environment is one of the EM functions in biological control measures. Harmful insects can potentially be repelled by the odor emitted by EM as it serves as a prophylactic spray. EM-5 and EMF. P.E. are good prophylactics that repel insects and are not toxic to frogs, ladybirds, dragonflies, and spiders, among others. However, effective microorganisms do attract fruit flies and can sterilize the females [35, 36]. By increasing the antagonistic activities and competition of the microbes in EM inoculants, pathogens and pests are controlled and/or suppressed by these natural processes [3, 35, 36].

3.7 Benefits of applying EM

Some beneficial impacts of EM in agricultural production are as follows:

- Enhancement of the chemical, biological, and physical environment of the soil (improve organic matter fertilizer efficacy) and suppression of pests as well as soil-borne pathogens

- Enhancement of flowering, germination of seeds, ripening, and fruiting in plants

- Increased yield in crops and improvement of crop photosynthetic capacity

- Crop quality and yield are thus enhanced, resulting from the above-mentioned beneficial effects of EM [3].

3.7.1 Effects of EM on organic matter

Soil biological, chemical, and physical characteristics can be improved by the addition of organic manures in addition to the supply of multiple nutrients. Farmers prefer mineral fertilizers in comparison with organic options because mineral fertilizers are quickly released and become available to the plant almost immediately. Organic fertilizers, on the other hand, are slowly released over a long period, and this may not quickly lead to increased yield. The release of nutrients to plants is thought to be stimulated by combining organic manures with EM. The improvement of soil and crop quality after inoculating agro-systems with EM has been shown [46]. The rapid growth of the plant results from the rapid mineralization of organic materials following EM application into the soil. This is because there is a rapid increase in beneficial microorganisms that consequently aid plant growth [3]. The application of EM or organic materials alone has not been shown to significantly increase yield [47]. However, their integrated use resulted in a 44% increase in yield over the control. Application of EM with mineral fertilizer in this case resulted in a slight increase in yield (14%) over the mineral fertilizer alone, demonstrating that EM is more effective when applied with organic manures. EM is comprised of various microbes that respond well when sufficient organic matter is available. This is why mineral fertilizers exhibited a slower release in comparison to EM. Yaseen et al. [48] revealed that arbuscular mycorrhizal and Rhizobia and inoculation of bean plants significantly increased pod yield in organic matter-supplemented plots compared to those treated with synthetic fertilizers. The impact of combining organic matter and EM was shown through a 38% increase in nitrogen leaf composition in comparison to the 16% observed in organic matter application alone [47]. Rapid mineral nutrient release into the soil for exploitation by plants is enhanced by EM through determination and stimulation of mineralization of organic materials [46]. In tomatoes, a higher phosphorus composition, 50 days after transplanting, was observed following the application of EM [49]. But, 90 days post-transplanting tomatoes, P and N contents were low in soils treated with EM, and this was potentially because of nutrients being absorbed by plants that showed faster growth and consequently higher yields. The release of P and N from organic matter-amended soils over an incubation period of 21 days at 60°C has been previously studied, and it was shown that the addition of EM increased both P and N in comparison to the control [50].

3.7.2 Effects of EM on the photosynthetic capacity of crops

The effects of combining EM with bokash on yield, photosynthesis, and plant growth in comparison with mineral fertilizers have been studied [51]. Dry matter yields during early growth phases but lower during later phases were observed by these authors but in mineral fertilizer-treated plants compared to EM plants. Vigorous growth, elevated photosynthesis, and increased root mass and activity throughout all growth stages were observed in EM and bokash-treated plants compared to mineral fertilizer-treated plants. Bokash-EM-treated plants lead to well-developed roots, and these are essential in maintaining increased photosynthetic activities and growth rates [52]. EM mineralizes nutrients from bokash and therefore ensures sustained availability of nutrients and consequently higher plant growth rates [53]. Senescence delay and root activity stimulation are potentially stimulated by growth regulators in EM [52]. Plant growth regulators such as abscisic acid, auxins, and gibberellins play a critical role in the development and growth of plant roots [54]. Metabolism and plant growth can be enhanced by some bioactive compounds that are produced by actinomycetes, fungi, and/or bacteria [51]. However, the stimulation of plant metabolic processes and growth by EM is not yet fully understood. However, the ability of EM to biosynthesize antioxidants is thought to be responsible for these microbes' beneficial effects [3].

3.7.3 Effects of EM on crop yield

Soil-borne pathogen suppression, an increase in plant nutrients, and an increase in yield are all enhanced by the application of EM [3]. First-grade sweet corn, peas, and onions are enhanced by the application of EM and molasses [3].

4. Organic manures

Traditional agricultural practices include applying organic manures at agronomic rates for plant nutrition delivery and advantageous impacts on physical qualities [55]. Due to a growing interest in the disposal of the enormous volumes of organic waste being generated over the past few decades, the impacts of organic manures on soil qualities have attracted new attention.

4.1 Effects of organic manures on physical and chemical properties of soil

According to Liu [56], organic manure has an impact on the soil's bulk density, stability, aggregation, pH, buffer capacity, cation exchange capacity, soil encrustation, water infiltration, soil penetrability, moisture content, drainage, tilth, aeration, soil temperature, as well as the supply and availability of nutrients for plant growth.

4.1.1 Soil bulk density

The physical qualities of the soil are improved by the addition of organic manures because it increases the amount of organic matter in the soil and the activity of its microbial population [57]. As a result of net increases in soil organic C, there is a direct correlation between changes in bulk density and water-holding capacity in the soil [58]. Because the denser mineral portion of the soil is diluted by the addition of organic manures, the bulk density of the soil falls. Organic manures have larger and more numerous pores and are less thick [59]. The humic components of organic manures reduce the flexibility, cohesiveness, and stickiness of clayey soils, causing the development of stable aggregates in the soil [60]. After the lag phase that occurs after adding organic manures to soils, the amount of microbial biomass increases. The physical entanglement of fungal hyphae and the synthesis of extra-cellular polysaccharides, which connect soil aggregates and hence boost aggregate stability, occur in conjunction with an increase in the microbial biomass pool [55]. In comparison to fresh organic manures, composted organic manures cause a slower, more gradual rise in aggregate stability [57].

4.1.2 Soil pH and cation exchange capacity

Cations like calcium, potassium, and magnesium are held in exchangeable forms by humus colloids so that they can be used by plants and are not washed away by water. According to Ampong et al. [60], "Organic matter also provides much of the pH buffering capacity in soils through its cation exchange capacity and acid and base functional groups." According to Fang et al. [59], applying organic manure caused the pH of the soil where composted organic materials had been mixed into acidic soil to rise. However, neutral or basic soils did not yield the same findings. Masmoudi et al. [61] demonstrated the high CEC of soils treated with organic manures. When compared to clay soil, the CEC on sandy soil rose by a ratio of 5–10.

4.1.3 Soil water content and soil water holding capacity

By increasing infiltration capacity and hydraulic activities, the addition of organic manures to the soil reduces surface crusting, reduces soil particle displacement by raindrops, reduces runoff water, reduces water lost through evaporation, improves drainage, and enhances root penetration [57]. By adding organic manures to soils,

soils can hold more water, which is regulated by the number, size, and distribution of pores. At low tensions, an increase in the quantity of tiny pores is principally responsible for an increase in water-holding capacity. The amount of surface area and the thickness of the water films on these surfaces dictate the soil moisture content at higher pressures where practically all the pores are air-filled [58]. The addition of organic manures to soils also expands the soil's specific surface area and boosts its ability to hold onto water under higher pressures [62].

4.1.4 Soil nutrients

Micronutrients, phosphorus, sulfur, nitrogen, and other elements are contained in organic manures and are gradually released through mineralization. The degree to which organic manures are mineralized is greatly influenced by their quality. As a result, N mineralization and immobilization, which, in turn, are regulated by C: N, lignin: N, polyphenol: N, and (lignin + polyphenol): N ratios, lignin and polyphenols, N% of the organic manure used, and moderate N release from organic materials [63]. In the early stages of decomposition, mineralization or immobilization processes will predominate depending on the chemical makeup of the organic manures, particularly the C: N ratio. The rate of material decomposition and subsequent soil turnover of decomposed C and N determine how much inorganic N is released from organic manures into the soil [63]. According to Ndambi et al. [64], the amount, kind, and duration of manure utilized all affect how successful it is as a source of nutrients for plants. Long-term addition of low-quality organic inputs can boost soil organic C build-up without necessarily boosting output. For instance, cropping system research in India found that using wheat straw and urea together significantly lowered yields, but using Sesbania green manure and urea together increased yields when compared to using urea alone [65]. According to Adekiya et al. [57], low-quality organic manures have low concentrations of soluble C and N, which are necessary to boost the activities of the soil's microbial pool and, thus, reduce crop output. Numerous crop wastes and animal manures can temporarily immobilize nitrogen (N) and have low-quality nitrogen contents that vary from 1.8 to 2.0%. Several crops have been produced using fresh organic manures as well as composted organic manures as a source of nutrients, with varied degrees of success. The results of adding organic manure to a soil depend on the quality of the manure and the characteristics of the soil. Manure application rates of 5.5 to 11 t ha^{-1} for field crops and double that amount for plants needing more nutrients have been recommended for South Africa [66]. The precise amount of organic manure required for optimal crop production is difficult to predict since it depends on the type of manure used, the type of soil, the needs of the crop, and the current environmental circumstances. Small-holder farmers in South Africa's Eastern Cape Province apply manure at rates that range from 0.3 to 18.2 t ha^{-1} [67]. However, manure is applied at rates of 25–100 t ha^{-1} [68]. Manure may be harmful in high concentrations to humans, animals, and plants [69]; hence, care must be taken to limit any negative consequences. According to Reimer et al. [70], 100 t ha^{-1} of compost added annually is sufficient to alter the physical properties of the soil and boost output. Numerous studies have shown that applying both organic manures and inorganic mineral sources together has a greater positive impact on growth and production than applying either one alone. This can be because organic manures do not have enough of all the necessary plant nutrients. When inorganic fertilizers are mixed with organic manures, their soluble minerals will mix with the organic portion because inorganic fertilizers are quickly leached. The nutrients will then be gradually released over time in the form of microbial pool by-products and under the influence of organic acids [57].

4.2 Types of organic manures

You can use organic manures on soil as compost or right out of the bag. Fresh organic materials have higher net N mineralization rates and higher inorganic N concentrations than composted manure, according to Cambardella et al. [71]. According to Paul and Beauchamp's study from 1994, plants treated with fresh organic manure showed more dry matter in their first growth season than plants treated with composted manure.

4.2.1 Compost

As organic materials are broken down, decomposed, and stabilized by natural microorganisms in a moist, warm, aerobic environment, carbon dioxide, water, minerals, and stabilized organic matter are produced, and pathogenic microbes are eliminated by enzymatic combustion and the heat generated [72]. Organic waste can be recycled by composting and used as a soil amendment. Composting organic waste has been shown to improve soil fertility, soil structure, and plant growth. According to Cambardella et al. [71], applying un-composted trash or compost that has not been stabilized to soil might cause phytotoxicity and nutrient immobilization.

4.2.2 Animal manures

Animal manure is used in Africa to improve soil fertility, and its advantages are well known. Animal food modifications, collection methods, and storage methods all affect the nutrient content of animal dung. Before the development of mineral fertilizers, manure and other livestock waste products were the only methods of increasing soil productivity. Animal manure application is a frequent practice in Zimbabwe, and it has been discovered that the quality of manure as a source of plant nutrients varies greatly in terms of chemical composition [73]. The most frequent manures used in southern Africa are those produced by goats, sheep, cattle, and chicken, with cattle producing two-thirds of the total amount of manure utilized and sheep and goats producing the remaining third. Species, animal diet, mineral particle content, and storage conditions all have a significant impact on the nutritional makeup of animal manures [66]. However, in the Eastern Cape Province, the nutrients in animal manures fall within the ranges recorded for manures in West African nations and range from 9.9 to 16.7 g N, 2.0 to 3.6 g P, and 17.2 to23.7 g K kg^{-1} [68]. The manures have a low P content, though. The differences in animal diet that affect how nitrogen is divided between feces and urine cause the nutrient content of manures to vary. When animals are fed high-quality food, N is lost by volatilization and expelled through urine [74]. The amount of nitrogen (N) excreted in feces is thought to rise when diets with a significant number of tannins are consumed. N in manures from animals given a diet high in tannin is particularly resistant to mineralization in the soil, according to a study by Halvorson et al. [75]. According to Mkile's [68] analysis of the nutritional content of various manures in the Eastern Cape region of South Africa, goat manure had the highest levels of N, P, and K, whereas sheep dung had the lowest levels.

5. Conclusion

Through the rapid proliferating of their constituents, more advantageous microorganisms, and subsequent suppression of soil-borne harmful organisms, the application of various brands of EM has been demonstrated to boost crop growth and quality, lowering the occurrence of pests and diseases. They may also effectively mineralize organic soil matter, which enhances nutrient availability and is how EM might boost soil health and crop growth. It has been well documented that EM has positive benefits on agricultural output by promoting plant germination, blooming, fruiting, and ripening. However, several studies have not consistently demonstrated the positive effects of EM on yield, indicating that the findings are not repeatable.

Additional information

"Parts of this chapter were previously published in the masters dissertation by the same author: Lindani Ncube. Evaluation of the effective micro-organisms (EM) on soil chemical properties and yield of selected vegetables in the Eastern Cape, South Africa. 2008. [Unpublished MSc Dissertation. University of Fort Hare. Available from: http://vital.seals.ac.za:8080/vital/access/manager/Repository/vital:11886?site_name=GlobalView&view=null&f0=sm_subject%3A%22Microorganisms+--+South+Africa+--+Eastern+Cape%22&sort=null"

Author details

Lindani Ncube
Department of Environmental Sciences, College of Agriculture and Environmental Sciences, UNISA, Science Campus, Florida, Johannesburg, South Africa

*Address all correspondence to: ncubel@unisa.ac.za

References

[1] Bvenura C, Afolayan AJ. Mineral uptake in *Solanum nigrum* L. cultivated on fertilizer amended soils of the Eastern Cape, South Africa. Journal of Applied Botany and Food Quality. 2014;**87**:220-226

[2] Kamara A, Conteh A, Rhodes ER, Cooke RA. The relevance of smallholder farming to African agricultural growth and development. African Journal of Food Agriculture, Nutrion and Development. 2019;**19**(1):14043-14065

[3] Safwat SM, Matta ME. Environmental applications of effective microorganisms: A review of current knowledge and recommendations for future directions. Journal of Engineering and Applied Science. 2021;**68**:48

[4] Ortiz A, Sansinenea E. The role of beneficial microorganisms in soil quality and plant health. Sustainability. 2022;**14**:5358

[5] UNDESA. United nations department of economic and social affairs - population, food security, nutrition, and sustainable development. Policy Brief. 2021;**No. 102**

[6] Ritchie H, Rosado P, Roser M. Agricultural Production. 2023. Published Online at OurWorldInData.Org. Available from: https://ourworldindata.org/agricultural-production

[7] Kuyah S, Sileshi GW, Nkurunziza L, Chirinda N, Ndayisaba PC, Dimobe K, et al. Innovative agronomic practices for sustainable intensification in sub-Saharan Africa. A Review. Agronomy for Sustainable Development. 2021;**41**:16

[8] Gehrels R, Garrett E. Chapter 11—Rising sea levels as an indicator of global change. In: Climate Change, Observed Impacts on Planet Earth. 3rd ed. Vol. 2021. Amsterdam, The Netherlands: Elsevier; 2021. pp. 205-217

[9] FAO/IFAD/UNICEF/WFP/WHO. The state of food security and nutrition in the world 2021. In: Transforming Food Systems for Food Security, Improved Nutrition and Affordable Healthy Diets for All. Rome, Italy: FAO; 2021. p. 2021

[10] The World Bank. 2021. Global Wealth Has Grown, but at the Expense of Future Prosperity: World Bank. Press Release No: 2022/23/ENR

[11] FAO. World Food and Agriculture—Statistical Yearbook 2021. Rome, Italy: FAO; 2021

[12] STATISTA. Global Production of Vegetables in 2021, by Type. 2023. Available from: https://www.statista.com/statistics/264065/global-production-of-vegetables-by-type/

[13] Bvenura C, Sivakumar D. The role of wild fruits and vegetables in delivering a healthy and balanced diet. Food Research International. 2017;**99**:15-30

[14] Struik PC, Kuyper TW. Sustainable intensification in agriculture: The richer shade of green. A review. Agronomy and Sustainable Development. 2017;**37**:39

[15] Binswanger-Mkhize HP, Savastano S. Agricultural intensification: The status in six African countries. Policy. 2017;**67**:26-40

[16] Stewart ZP, Pierzynski GM, Middendorf BJ, Prasad PVV. Approaches to improve soil fertility in sub-Saharan Africa. Journal of Experimental Botany. 2020;**71**(2):632-641

[17] OBG – Oxford Business Group. Agriculture in Africa 2021. 2021. Available from: https://oxfordbusinessgroup.com/wp-content/uploads/files/blog/specialreports/960469/OCP_Agriculture_Africa_Report_2021.pdf

[18] Gardiner RKA, Mabogunje AL. Agriculture of Africa. Britanica. 2023. Available from: https://www.britannica.com/place/Africa/Agriculture

[19] Demay J, Ringeval B, Pellerin S, Nesme P. Half of global agricultural soil phosphorus fertility derived from anthropogenic sources. Nature Geoscience. 2023;**16**:69-74

[20] Tetteh FM. Challenges to soil protection and sustainable Management in Africa. In: Ginzky H et al., editors. International Yearbook of Soil Law and Policy. Cham: Springer; 2021

[21] Singh M, Maharjan KL. Sustainability of farming system: An overview. In: Sustainability of Organic Farming in Nepal. Singapore: Springer; 2017

[22] Zhong R, Pei F, Yang K, Xia Y, Wang H, Yan G. Coordinating socio-economic and environmental dimensions to evaluate regional sustainability—Towards an integrative framework. Ecological Indicators. 2021;**130**:108085

[23] Gamage A, Gangahagedara R, Gamage J, Jayasinghe N, Kodikara N, Suraweera P, et al. Role of organic farming for achieving sustainability in agriculture. Farming System. 2023;**100005**:1-14

[24] Hartmann M, Six J. Soil structure and microbiome functions in agroecosystems. Nature Reviews Earth and Environment. 2023;**4**:4-18

[25] FAO/UN. The World of Organic Agriculture 2021-Report. Rome; 2021

[26] Röös E, Mie A, Wivstad M, Salomon E, Johansson B, Gunnarsson S, et al. Risks and opportunities of increasing yields in organic farming. A review. Agronomy for Sustainable Development. 2018;**38**:1-21

[27] Tuomisto HL, Hodge ID, Riordan P, Macdonald DW. Does organic farming reduce environmental impacts?--a meta-analysis of European research. Journal of Environmental Management. 2012;**112**:309-320

[28] Aulakh CS, Sharma S, Thakur M, Kaur P. A review of the influences of organic farming on soil quality, crop productivity and produce quality. Journal of Plant Nutrition. 2022;**45**:1884-1905

[29] Williams DM, Blanco-Canqui H, Francis C, Galusha TD. Organic farming and soil physical properties: An assessment after 40 years. Agronomy Journal. 2017;**109**:600-609

[30] Rawat L, Bisht TS, Naithani DC. Plant disease Management in Organic Farming System: Strategies and challenges. In: Singh KP, Jahagirdar S, Sarma BK, editors. Emerging Trends in Plant Pathology. Singapore: Springer; 2021

[31] Xing TT, Cai A, Lu C, Ye H, Wu H, Huai S, et al. Increasing soil microbial biomass nitrogen in crop rotation systems by improving nitrogen resources under nitrogen application. Journal of Integrative Agriculture. 2022;**21**(5):1488-1500

[32] Wilhelm RC, Singh R, Eltis LD, Mohn WW. Bacterial contributions to delignification and lignocellulose degradation in forest soils with metagenomic and quantitative stable isotope probing. The ISME Journal. 2019;**13**(2):413-429

[33] Prescott LM, Harley JP, Klein DA. Microbiology. 3rd ed. New York, USA: The McGraw-Hill Companies, Inc; 2002

[34] Nayak NP, Sar K, Sahoo BK, Mahapatra P. Beneficial effect of effective microorganism on crop and soil- a review. Journal of Pharmacognosy and Phytochemistry. 2020;**9**:3070-3074

[35] Anonymous. Organic farming enters the mainstream. Nature (London). 2004;**428**:783

[36] Anonymous. EM user's Manual. 70 Lloys Ellis Ave, Mnandi, Centurion: EMROSA, (PTY) LTD; 2005

[37] Ke B, Liang YF, Zhong ZX, Higa T, Aruoma OI. Evaluation of the toxicity and safety of the antioxidant beverage effective microorganisms-X (EM-X) in animal models. Environmental Toxicology and Pharmacology. 2005;**20**(2):313-320

[38] Olle M, Williams IH. Effective microorganisms and their influence on vegetable production—A review. The Journal of Horticultural Science and Biotechnology. 2013;**88**:380-386

[39] Aly MZ, Zaki MA, Mansour AT, Srour TM, Omar EA. Effect of effective microorganisms (Em) and molasses, wheat bran, and their mixture in a biofloc system on microbial protein production, water quality, growth performance and feed utilization of Nile tilapia (*Oreochromis niloticus*) fingerlings. Journal of Animal and Poultry Production. 2017;**8**(11):443-449

[40] Zu Schweinsberg-Mickan MS, Müller T. Impact of effective microorganisms and other biofertilizers on soilmicrobial charakteristics, organic-matter decomposition and plant growth. Journal of Plant Nutrition and Soil Science. 2009;**172**:704-712

[41] Schütz L, Gattinger A, Meier M, Müller A, Boller T, Mäder P, et al. Improving crop yield and nutrient use efficiency via biofertilization—A global meta-analysis. Frontiers in Plant Science. 2018;**8**:2204

[42] Iriti M, Scarafoni A, Pierce S, Castorina G, Vitalini AND, S. Soil application of effective microorganisms (EM) maintains leaf photosynthetic efficiency, increases seed yield and quality traits of bean (*Phaseolus vulgaris* L.) plants grown on different substrates. International Journal of Molecular Sciences. 2019;**20**:2327

[43] Aliyu IA, Yusuf AA, Atta A. Evaluation of indigenous rhizobial isolates in search for candidate strain for commercial production. Bayero Journal of Pure and Applied Sciences. 2019;**11**:33

[44] Chavoshi S, Nourmohamadi G, Madani H, Abad HHS, Fazel MA. The effects of biofertilizers on physiological traits and biomass accumulation of red beans (Phaseolus vulgaris cv.Goli) under water stress. Iranian Journal of Plant Physiology. 2018;**8**:2555-2562

[45] Santos MS, Nogueira MA, Hungria M. Microbial inoculants: Reviewing the past, discussing the present and previewing an outstanding future for the use of beneficial bacteria in agriculture. AMB Express. 2019;**9**:1-13

[46] Joshi H, Duttand S, Choudhary P, Mundra SL. Role of effective microorganisms (EM) in sustainable agriculture. International Journal of Current Microbiology and Applied Sciences. 2019;**8**(3):172-181

[47] Khaliq A, Abbas MK, Hussain T. Effects of integrated use of organic and inorganic nutrient sources with effective micro-organisms (EM) on seed cotton yield in Pakistan. Bioresource Technology. 2006;**97**:967-972

[48] Yaseen T, Ali K, Munsif F, Rab A, Ahmad M, Israr M, et al. Influence of

arbuscular mycorrhizal fungi, rhizobium inoculation and rock phosphate on growth and quality of lentil. Pakistan Journal of Botany. 2016;**48**(5):2101-2107

[49] Xu HL. Effects of microbial inoculant and organic fertilizers on the growth, photosynthesis, and yield of sweet corn. Crop Production. 2000;**3**:183-214

[50] Piyadasa ER, Attanayake KB, Ratnayake ADA, Sangakkara UR. The role of effective microorganisms in releasing nutrients from organic matter. In: Second Conference on Effective Microorganism (EM). Saraburi, Thailand: Kyusei Nature Farming Center; 1995. pp. 7-14

[51] Mukhtar NK, Azhar MS, Bakar TH, Md Zain N, Abu Bakar TH, Md Zain N. Assessing the effects of microorganisms (EM) and Bokashi on okra (*Abelmoschus esculentus* L. Moench) towards plant growth and yield performance. In: IOP Conference Series: Earth and Environmental Science. 2022. pp. 1-8, 1102-012055

[52] Shin K, Diepen GV, Blok WJ, Bruggen AV. Variability of effective micro-organisms (EM) in bokashi and soil and effects on soil-borne plant pathogens. Crop Protection. 2017;**99**:168-176

[53] Quiroz M, Céspedes C. Bokashi as an amendment and source of nitrogen in sustainable agricultural systems: A review. Journal of Soil Science and Plant Nutrition. 2019;**19**:237-248

[54] Emenecker RJ, Strader LC. Auxin-Abscisic acid interactions in plant growth and development. Biomolecules. 2020;**10**(2):281

[55] Ali Q, Ashraf S, Kamran M, Ijaz M, Rehman A, Tahir M, et al. Organic manuring for agronomic crops. Agronomic crops. In: Hasanuzzaman M, editor. Agronomic Crops. Singapore: Springer; 2019

[56] Liu S, Wang J, Pu S, Blagodatskaya E, Kuzyakov Y, Razavi BS. Impact of manure on soil biochemical properties: A global synthesis. The Science of the Total Environment. 2020;**745**:141003

[57] Adekiya AO, Ejue WS, Olayanju A, Dunsin O, Aboyeji CM, Aremu C, et al. Different organic manure sources and NPK fertilizer on soil chemical properties, growth, yield, and quality of okra. Scientific Reports. 2020;**10**(1):16083

[58] Jiao S, Li J, Li Y, Xu Z, Kong B, Li Y, et al. Variation of soil organic carbon and physical properties in relation to land uses in the Yellow River Delta, China. Scientific Reports. 2020;**10**(1):20317

[59] Fang H, Liu K, Li D, Peng X, Zhang W, Zhou H. Long-term effects of inorganic fertilizers and organic manures on the structure of a paddy soil. Soil & Tillage Research. 2021;**213**:105137

[60] Ampong K, Thilakaranthna MS, Gorim LY. Understanding the role of humic acids on crop performance and soil health. Frontiers in Agronomy. 2022;**4**:848621

[61] Masmoudi S, Magdich S, Rigane H, Medhioub K, Rebai A, Ammar E. Effects of compost and manure application rate on the soil physico-chemical layers properties and plant productivity. Waste Biomass Valor. 2020;**11**:1883-1894

[62] Choudhary RC, Bairwa HL, Kumar U, Javed T, Asad M, Lal K, et al. Influence of organic manures on soil nutrient content, microbial population, yield and quality parameters of pomegranate (Punica granatum L.) cv. Bhagwa. PloS one. 2022;**17**(4):e0266675

[63] Pansu M, Thuries L. Kinetics of C and N mineralization, N immobilization and N volatilization of organic inputs in soil. Soil Biology & Biochemistry. 2003;**35**(1):37-48

[64] Ndambi OA, Pelster DE, Owino JO, de Buisonjé F, Vellinga TV. Manure management practices and policies in sub-Saharan Africa: Implications on manure quality as a fertilizer. Frontiers in Sustainable Food Systems. 2019;**3**:29

[65] Singh G, Jalota SK, Singh Y. Manuring and residue management effects on physical properties of a soil under the rice–wheat system in Punjab, India. Soil & Tillage Research. 2007;**94**:229-238

[66] Okorogbona AOM, Adebisi LO. Animal manure for smallholder agriculture in South Africa. In: Lichtfouse E, editor. Farming for Food and Water Security. Sustainable Agriculture Reviews. Vol. 10. Dordrecht: Springer; 2012

[67] Yoganathan S, Sotana MM, van Averbeke W, Mandiringana OT, Materechera S, Harris PJC, et al. Kraal manure as a fertilizer for small-scale crop production in central eastern cape, South Africa. In: Proceedings of AFSR-E 15th International Symposium, 29 November–4 December. Pretoria, South Africa: University of Pretoria; 1998. pp. 361-368

[68] Mkile Z. The use and agronomic effectiveness of kraal manures in the Transkei region of the Eastern Cape, South Africa [unpublished M Sc. Agric, dissertation]. Alice, South Africa: University of Fort Hare; 2001

[69] Rayne N, Aula L. Livestock manure and the impacts on soil health: A review. Soil Systems. 2020;**4**:64

[70] Reimer M, Kopp C, Hartmann TE, Zimmermann H, Ruser R, Schulz R, et al. Assessing long term effects of compost fertilization on soil fertility and nitrogen mineralization rate. Journal of Plant Nutrition and Soil Science. 2023;**186**(2):217-233

[71] Cambardella CA, Richard TL, Russell A. Compost mineralization in soil as a function of composting process conditions. European Journal of Soil Biology. 2003;**39**:117-127

[72] Ouatmane A, Provenzano MR, Hafidi M, Senesi N. Compost maturity assessment using colorimetry, spectroscopy and chemical analysis. Compost Science and Utilization. 2000;**8**:124-134

[73] Parwada C, Chigiya V, Ngezimana W, Chipomho J, Bandason W, Nyamushamba GB. Use and Management of Animal Manure by the communal farmers, Seke District, Mashonaland East Province, Zimbabwe. Research on World Agricultural Economy. 2021;**2**(1):6-15

[74] Uwizeye A, de Boer IJ, Opio CI, Schulte RP, Falcucci A, Tempio G, et al. Nitrogen emissions along global livestock supply chains. Nature Food. 2020;**1**:437-446

[75] Halvorson JJ, Kronberg SL, Hagerman AE. Effects of dietary tannins on total and extractable nutrients from manure. Journal of Animal Science. 2017;**95**(8):3654-3665

Chapter 6

Movement of Water and Solutes in Agricultural Soils

Thaís Nascimento Pessoa, Aline Martineli Batista, Rodrigo Nogueira de Sousa and Paulo Leonel Libardi

Abstract

Intensive agricultural systems have increased in recent decades. These systems usually provide high soil disturbance that promotes negative consequences on soil physical, chemical, and biological processes. Soil compaction degrades soil structure causing changes in soil porosity, pore size distribution, and pore space geometry. Pore space dynamics affects the movement of water and the transport of solutes in soils because both are highly dependent on the properties of this space. Agricultural activities, which include the use of mechanization, soil fertilizers, herbicides, and pesticides with the aim of creating a favorable condition for plant growth, can favor the soil erosion processes and increase the transport of contaminants to groundwater. Models for predicting water movement and transport of solute in soils are important to understand different scenarios of groundwater contamination and to improve appropriate soil management. Therefore, negative consequences and environmental impacts related to agricultural practices should be studied.

Keywords: hydraulic properties, available water content, groundwater contamination, soil structure, modeling of soil processes, land-use change

1. Introduction

The intensification of agricultural land use is of great concern mainly due to its environmental impacts in terms of chemical, physical, and biological degradation (e.g., soil loss, faster mineralization of organic matter, biological disturbance, and chemical pollution) [1]. When agrochemicals, such as fertilizers, insecticides, herbicides, and other pesticides, are applied in a cropping to improve plant growth, their organic and inorganic molecules can become readily available in the soil solution through rainwater or irrigation. In other words, since the transport of solutes of the solution in soil is the primary vehicle for groundwater contamination [2], the main driver of water pollution is the migration of contaminated solution from topsoil to groundwater.

Transport processes depend on inherent soil properties and are influenced by crop management practices, including crop rotation and tillage [3]. The adoption of agricultural practices such as crop rotations can improve soil quality, and the choice of cover crops plays an important role in soil properties. Thus, the adoption of different crop rotations will have different effects on soil aggregation [4]. On the other hand, the formation of soil aggregates is responsible for determining the soil structure and consequently the pore network [5]. Therefore, changes are also observed in the water and solutes movement in the soil. The roots effect on the soil structure varies among plants [6], and apparently the main factor that reflects in the movement of water and consequently dissolved solutes in the soil is the root exudation of organic substances, mainly mucilage [7–9], a polymeric gel released by most plant roots [10]. In addition, the formation of galleries or channels in the soil due to root decomposition also plays an important role in this process. These channels can modify soil porosity and soil permeability, further affecting the dynamics of water flow and nutrient transport within the soil.

Soil structure defines the pore geometry that controls soil water dynamics. Water movement and solute transport in soils are also influenced by soil aggregation, pore volume, pore size distribution, and pore connectivity [11]. Typically, field soils have different soil properties that vary spatially as a consequence of pedogenic processes [12].

The movement and retention of water in soil is very important in agriculture, and the physical attributes that reflect their intensities are the hydraulic conductivity and the water retention curve, respectively [13]. The degree of soil compaction typically decreases the saturated hydraulic conductivity as a result of the reduction in macropores [14, 15]. Soil hydraulic properties such as available water content and hydraulic conductivity can be used to assess the long-term effects of soil land use and management [16]. Bagnall et al. [17] highlighted field capacity as an interesting soil health indicator to evaluate the impact of soil management practices on hydraulic properties, mainly measured in undisturbed soil samples. In addition, soil bulk density, soil organic carbon, and aggregate stability can be important indicators that reflect the response of management practices.

2. Water flow in saturated and unsaturated soils

In 1855, Henry Darcy carried out an experiment on water movement in columns of sand under saturation and steady-state conditions and concluded that under these conditions, flow rate is (i) directly proportional to the difference between piezometric heads acting on the ends of the sand columns, (ii) directly proportional to the cross-sectional area of the sand columns, and (iii) inversely proportional to the length of the columns. The application of *Darcy's Law* is very common nowadays in all fields of science that involve studies about flow in saturated porous media [18].

In **Figure 1**, we can see a saturated uniform soil column scheme with a constant water head (h_1) at the top ($z = L$) and a constant water head (h_2) at the bottom ($z = 0$). H_1 is the piezometric head at the top and H_2 is the piezometric head at the bottom of the soil column, and water is flowing under steady-state conditions.

The Darcy equation obtained from the conclusions of his experiment is Eq. (1):

$$Q = K_{sat} A \frac{H_1 - H_2}{L} \quad (1)$$

where $Q = V_w/t$ is the water flow ($m^3\,s^{-1}$), V_w is the volume of water (m^3) measured in the time t (s), K_{sat} is the saturated soil hydraulic conductivity (ms^{-1}), A is the

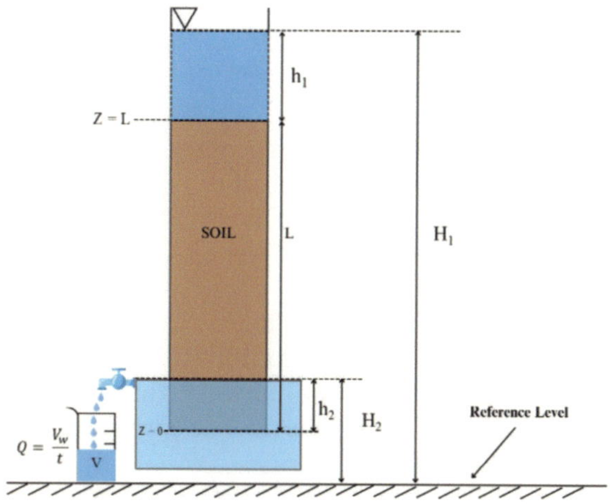

Figure 1.
A simplified diagram of Darcy's experiment.

soil cross sectional area (m^2), H_1-H_2 is the difference in piezometric head (m), and L is the length of the column (m).

The equipment most commonly used in soil physics laboratories to determine K_{sat} has been the constant water head permeameter in which Darcy's experiment (diagram in **Figure 1**) is developed, but using smaller soil columns with undisturbed (or disturbed) structure. Normally in the permeameter, we set $h_1 = h$ slightly greater than zero, $h_2 = 0$, and also $H_2 = 0$ (reference level coincident with column bottom soil surface) so that $H_1 = h + L$. With these fixed values for h_1, h_2, H_1, and H_2 in Eq. (1) and explaining K_{sat}, we get Eq. (2):

$$K_{sat} = \frac{V_w L}{At(h + L)} \tag{2}$$

Giving (Eq. (1)) a mathematical treatment and including the concept of water potentials to the Darcy equation (see item 3), it can be written as:

$$q = -K_{sat} \frac{d\psi_t}{dz} \tag{3}$$

where now $q = Q/A$ is the water flow density (m/s); ψ_t is the total potential (m), being equal to the sum of the gravitational potential ψ_g (m) and the pressure potential ψ_p (m); z is the vertical position coordinate (m); and $d\psi/dz$ is the total potential gradient; the negative sign has been included to indicate that when $q > 0$, the flow is in the positive direction of z (upwards), and when $q < 0$, the flow is in the negative direction of z (downwards).

Under unsaturated soil conditions, the development of the equation for water flow under steady-state conditions involved the work of Buckingham [19] and Richards [20] and is currently known as the Darcy–Buckingham equation (Eq. (4)):

$$q = -K(\theta)\frac{d\psi_t}{dz} \tag{4}$$

In this equation, K is a function of the water content θ and $\psi_t = \psi_g + \psi_m$, where ψ_m is the matric potential, which is also a function of θ.

Note that the Darcy–Buckingham equation is also valid for the saturation condition since, under this condition, just replace $K(\theta_s)$ by K_{sat} and ψ_m by ψ_p; θ_s is the water content in saturated soil.

The equation for water flow in soil under nonsteady-state or transient conditions (flow characteristics vary with time and position) is the Richards differential equation, which is the combination of the Darcy–Buckingham equation with the continuity equation. The continuity equation for the flow density of water in the soil in the vertical direction is Eq. (5) below:

$$\frac{d\theta}{dt} = -\frac{dq}{dz} \tag{5}$$

which combined with the Darcy–Buckingham equation also in the vertical direction results in the Richards equation (Eq. (6)):

$$\frac{d\theta}{dt} = \frac{d}{dz}\left[K(\theta)\frac{d\psi_t}{dz}\right] \tag{6}$$

The soil hydraulic conductivity that appears in equations of Darcy–Buckingham and Richards has been defined as the soil ability to transmit water [21].

Besides the constant water head method that is the own Darcy experiment to determine K_{sat} (see Eq. (3)), Deb and Shukla [22], in a complete bibliographic review, address about laboratory, field, correlation, and estimation methods for the determination of saturated and unsaturated soil hydraulic conductivity (**Figure 2**).

We believe it is worth noting that since it is difficult to measure the $K(\theta)$ function under field conditions is difficult, it can be calculated theoretically from data related to water retention in the soil, which is easier to measure [23–29]. In the Mualem-van Genuchten model [24, 25] from SWRC data, a relative soil hydraulic conductivity K_r as a function of soil water content θ, the $K_r(\theta)$ function, is calculated by the equation developed by van Genuchten [25] for this function, based on the work of Mualem [24], which has been called the Mualem-van Genuchten equation. To calculate $K_r(\theta)$ with this equation, there is only one parameter that needs to be known, which comes from the SWRC fitting equation used by van Genuchten [25]. $K(\theta)$ is then calculated by $K(\theta) = K_{sat} \times K_r(\theta)$, where K_{sat} can be determined, for example, by the constant water head permeameter.

Now, let's briefly discuss the importance of soil hydraulic conductivity.

Saturated hydraulic conductivity (K_{sat}), for instance, is very important for soil management in terms of soil erosion. Since soil structure and texture determine the geometry of the soil pores, they are essential factors influencing soil hydraulic conductivity [13]. Tillage and intensive use without conservation practices cause soil compaction and indirectly change soil hydraulic properties [30]. A decrease in soil

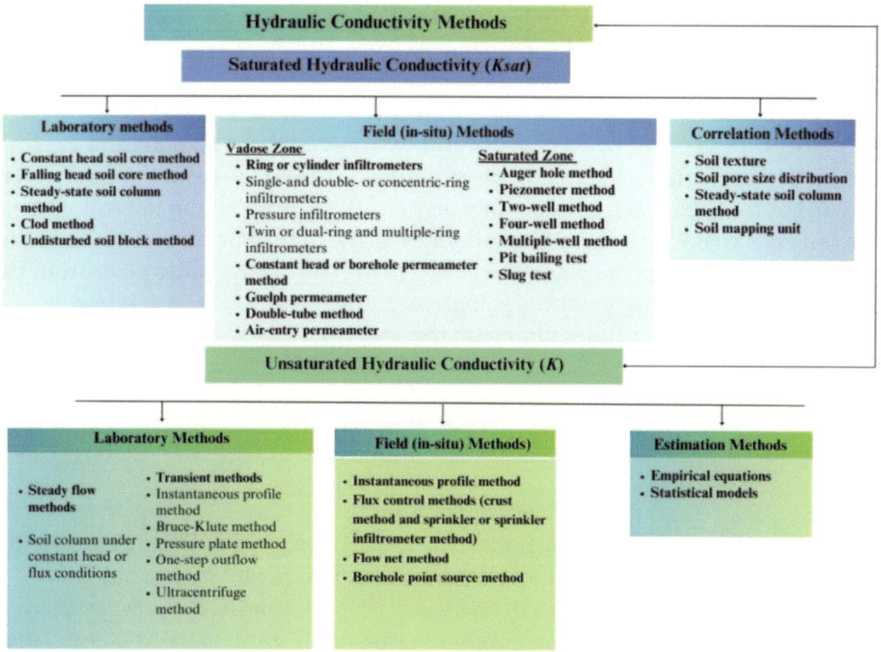

Figure 2.
Soil hydraulic conductivity determination methods. Source: Adapted from: Deb and Shukla [22].

porosity and an increase in soil bulk density are the main consequences of soil compaction: conducting pores (e.g., macropores and mesopores) are the most affected. A decrease in conducting pores and an increase in micropores (retention pores), in turn, decrease K_{sat} and change $K(\theta)$ function. Soil hydraulic conductivity is higher when the soil pore space is higher, when the soil is fractured (K_{sat} in this case), when the soil is well aggregated with better pore connectivity than when it is tightly compacted, and it also depends on the geometry of the pore space. Other aspects, such as changes in the composition of the exchange complex (e.g., the concentration of solutes) and the dispersion of clay particles clogging the pores, can reduce the conductivity [31]. Anyway, soil hydraulic conductivity under saturated and unsaturated conditions plays a fundamental role in estimating water and solutes movement and in studying soil-plant-water interactions [22].

Unsaturated K can be determined (i) in the laboratory using steady-flow methods that induce evaporation or infiltration (e.g., soil column under constant flow) or by transient methods (e.g., pressure plate method, by instantaneous profile method, and others); (ii) under field conditions, the instantaneous profile method and flux control methods; and (iii) finally by estimation methods using empirical equations or statistical models.

3. Water potentials and available water content in soil

Understanding the potential energy concepts of water in soil is crucial as they aid in predicting the potential movement of water in soil. The universal tendency of a body in nature is to move in response to its total potential energy (summation of the several potential energy components resulting from the several existing force fields acting in the body); i.e. the body moves from a position where its total potential energy is higher to a position where it is lower [32–34].

For convenience, in the case of water movement in the soil, it was defined the *total water potential in the soil* (ψ_t) as the difference between the total potential energy in the units of mass, volume, or weight of water inside the soil (ε) and the total potential energy in the units of mass, volume, or weight of the same water in the soil, but out of the soil (ε_0), that is, free from the influence of the matrix or free water also called *standard water.* Thus, the movement of water in the soil always takes place from the position where total water potential is highest to the position where it is lowest.

According to the force fields acting on the water in the soil, it follows that:

$$\psi_{\mathrm{t}} = \psi_{\mathrm{g}} + \psi_{\mathrm{p}} + \psi_{\mathrm{n}} + \psi_{\mathrm{m}} \quad (7)$$

where $\psi_g = \varepsilon_g - \varepsilon_g^0$ is the gravitational potential, $\psi_p = \varepsilon_P - \varepsilon_P^0$ is the pressure potential, $\psi_n = \varepsilon_n - \varepsilon_n^0$ is the pneumatic potential, and $\psi_m = \varepsilon_m - \varepsilon_m^0$ is the matric potential. ε_g, ε_P, ε_n, and ε_m are gravitational, pressure, pneumatic, and matric potential energies of the unit mass, volume, or weight of the water in the soil, respectively, and ε_g^0, ε_P^0, ε_n^0, and ε_m^0 are gravitational, pressure, pneumatic, and matric potential energies of the unit mass, volume, or weight of the standard water, respectively.

The matric potential has the greatest influence on the total potential under unsaturated soil conditions and it will be more emphasized in this chapter. The retention of water by the soil is controlled by the capillary and adsorption forces (i.e., components of the matric forces) (**Figure 3**).

Soil pore space heterogeneity does not allow to obtain a theoretical equation for the matric potential [29]. Consequently, devices were developed to establish a relation between matric potential and soil water content. The curve resulting from this relation is the soil water retention curve (SWRC) (**Figure 4**). **Figure 4** shows two examples of SWRC for clayey and sandy soils. SWRC can be used to estimate soil available water content, among other soil science applications. SWRC is affected by soil texture and structure. Soil structure is more important at higher matric potentials, while texture is more important at lower potentials. However, soil structure and texture together determine water retention and conduction throughout the SWRC. Using SWRC, it is possible to study the effects of compaction and management on soil structure (i.e., porosity, volume, distribution, and frequency of pores) and therefore on the soil physical-hydric properties.

There is a relationship between soil water content and water availability to plants. When the soil is saturated, water drains easily and is available to plants for a short period of time. As soil water content decreases and fast-draining macropores empty, the water-holding force of soil rises, decreasing considerably its downward movement. The soil reaches its field capacity (FC) at this stage, with water held only in micropores. The FC is typically referred to the "upper limit of soil water availability". The use of water by plants leads water to be also removed from the larger micropores, remaining in the smaller micropores and around the solid particles. When water is strongly retained by the soil matrix and the water uptake becomes very slow due to the high suction force demanded by the plants, the permanent wilting point (PWP) occurs. The PWP is also defined as the "lower limit of soil water availability" because the absorbed water content is no longer sufficient to maintain plant turgidity. The available water content is related to soil structure and texture. Clayey soils have a

DOI: http://dx.doi.org/10.5772/intechopen.114086

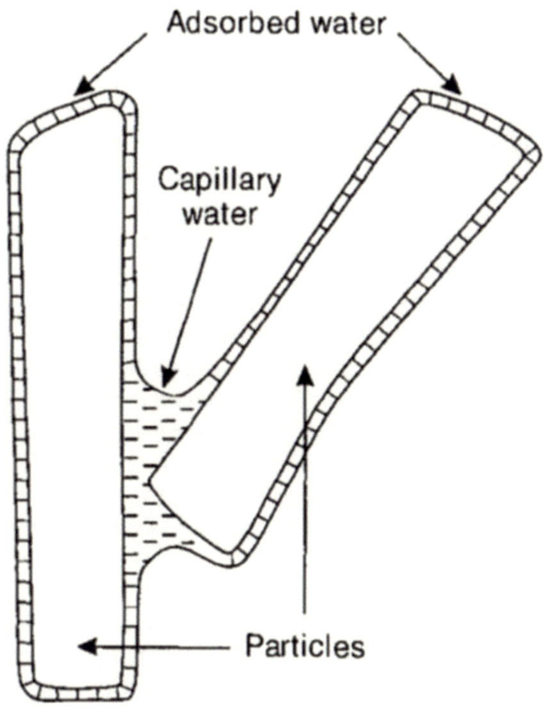

Figure 3.
Capillarity and adsorption phenomena that give rise to the matric potential. Source: Hillel [31].

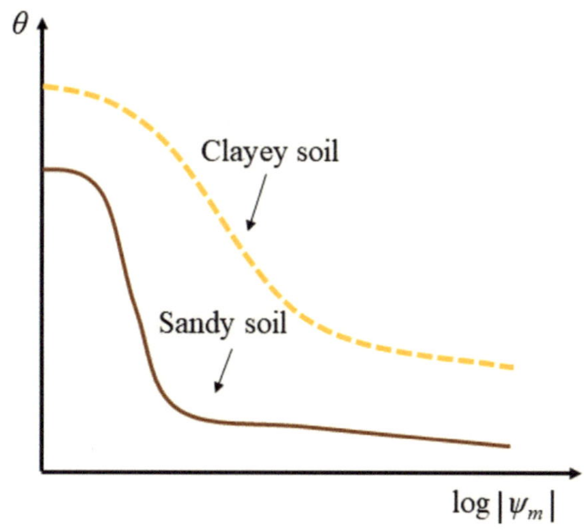

Figure 4.
Typical soil water retention curves (SWRCs) of clayey and sandy soils.

higher available water content than sandy soils because their greater specific surface area allows for greater water retention. Usually, clayey soils have a pore network with smaller pores (micropores) that are electronically charged and have a greater ability to hold water than sandy soils.

Soil organic matter (SOM) can also affect water availability due to its high water holding capacity. Consequently, an increase in SOM will result in an increase in the available water content. Costa et al. [35] observed higher FC and PWP in the soil surface layer due the higher SOM content in this layer. The effects of SOM on water availability are dependent on several factors. Lal [36] highlighted factors influencing this phenomenon, such as inherent soil properties, crop management, land use, and others (**Figure 5**).

Isolate measurements of soil physical properties should not always reflect the soil condition for plant growth and development [37]. Crop production can be limited by water and soil physical parameters [38]. Only when these limitations are removed can the difference between FC and PMP, i.e., the water available to plants, reflect the effect of soil structure on plant growth and development and crop production. However, this range of water content can be integrated into an indicator that overcomes the limitation of isolated measurements: the least limiting water range (LLWR) [39].

Initially, Letey [40] conceptualized and defined a range of non-limiting water contents for plant growth, called the non-limiting water range (NLWR). This parameter can be affected by mechanical resistance and/or aeration, especially in poorly structured soils with high bulk density, and is considered a way to relate soil physical properties to plant growth. It is based on the fact that mechanical resistance inhibits root growth, while roots require accessible water reservoirs and oxygen [38]. The NLWR was improved by Silva et al. [41] to quantify physical and structural soil quality. These authors used the term least limiting water range (LLWR) to describe this concept, which is an improvement over the NLWR. It consists of the range of soil water content in which limitations to plant growth are minimal and was considered an indicator of soil function for biomass production [42]. The LLWR can identify physical conditions that influence the functionality of physical, chemical, and biological processes [43].

The LLWR is a range of soil water contents that delimits critical soil physical limits for root growth (**Figure 6**) [38]. Basically, it is defined by an upper limit (wetter)

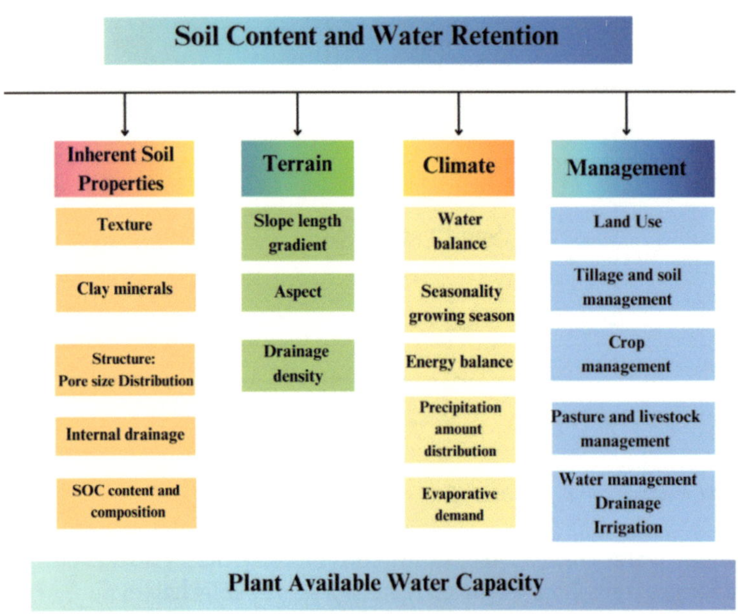

Figure 5.
Factors that influence soil water retention and available water content. Source: Lal [34].

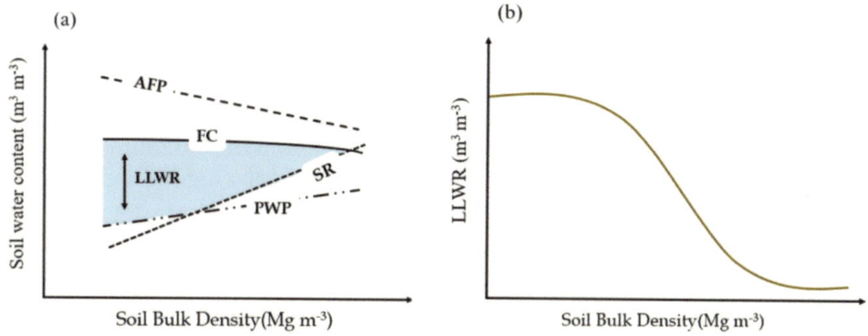

Figure 6.
The concept of least limiting water range (LLWR): (a) Soil water content as a function of soil bulk density based on changes in FC, PWP, and of soil water contents corresponding to air-filled porosity (AFP) and soil penetration resistance (SR); (b) LLWR as a function of soil bulk density. Source: Keller et al. [38].

and a lower limit (drier), where plant growth is not restricted [41]. The upper limit is considered to be the FC or the water content corresponding to 10–15% of soil aeration, while the lower limit is the PWP or the water content corresponding to a soil resistance (SR) value that limits root growth (varying between 2 and 4 MPa). However, other authors have made adjustments to the LLWR by replacing the PWP of the LLWR with the critical water content resulted in good results for both a perennial crop and crop rotations in Brazil [44, 45]. Recently, Silva et al. [46] added the critical soil water content for the crop at each phenological stage as the lowest limit of LLWR.

In a simple way, the LLWR allows the identification of physical properties that act as a limiting factor with increasing soil compaction as a function of bulk density (**Figure 6**). The LLWR can be used to evaluate the physical and structural quality of soils for crops grown in different soils and management systems and also in relation to processes related to nitrogen dynamics and soil carbon [39]. It is considered an effective index for influencing grain yield and for detecting the effects of tillage and cover crops [47, 48].

4. Movement of solutes in soils

Water and solute movement occur simultaneously in the soil. Understanding the processes of solute transfer in this environment, as well as the ionic interactions between the liquid and solid phases, is important for establishing management practices in the soil–plant–atmosphere system. Contaminants (e.g., heavy metals, salts, and organic molecules) and fertilizers can be rapidly transported by water flow into streams or soil layers where water is stored [49]. These substances are the main sources of groundwater and surface water contamination [50, 51]. To assess and reduce risks to human health, it is necessary to predict and know the mechanisms that influence the movement and fate of contaminants in runoff water [52]. Water flow and mass transfer in soils are very complex at the field scale due to the process of variation of fluid properties in the soil [53].

Solute movement through a porous medium is the process of displacement of ions in solution under unsaturated or saturated conditions of that medium [49]. In a porous medium, the mixture of two miscible fluids can be analyzed by different mathematical models [54]. Two processes dominate the flow of ions in soils: mass transfer and diffusion. Mass transfer refers to the movement of ions by the water flow, whereas diffusion process is the movement of ions due to gradient of activity of the ions [55].

Figure 7 shows a diagram of an experimental setup of a miscible displacement experiment. In this scheme, with the water already flowing in the column, when the solution of a solute of interest of concentration C_0 is placed in the column, it displaces the water; that is, in this case, the solute solution is the displacing fluid and the water is the displaced fluid. The study of solute movement is carried out using *Breakthrough Curves* (BTCs). BTCs are defined as the graph of C/C_0 (solute relative concentration) as a function of pore volume (Qt/V_0), where C is the concentration found in the effluent, C_0 is the initial concentration of the solute in the displacing fluid, Q is the volume of effluent collected per unit time, t is the time elapsed since the displacing liquid was added, and V_0 is the volume of the soil column occupied by the fluid [54–56].

Miscible displacement is the process that occurs when one liquid mixes with and displaces another liquid. The mixture occurs in the interface of the two fluids by diffusion and mass flow and depends on the fluid properties, such as viscosity and ionic concentration, and the material in which the fluids are contained. In the case of solute movement in soil, the properties are inherent to the soil. The solute transport process varies with the surface of the medium (adsorption and desorption processes, among others) and depends on the degree of saturation of the medium and the structure of the medium (e.g., pore size distribution, pore connectivity, and type of soil aggregates). Solute movement is the result of complex phenomena, and the interactions of this process can be analyzed by comparing measured data with simplified models [54].

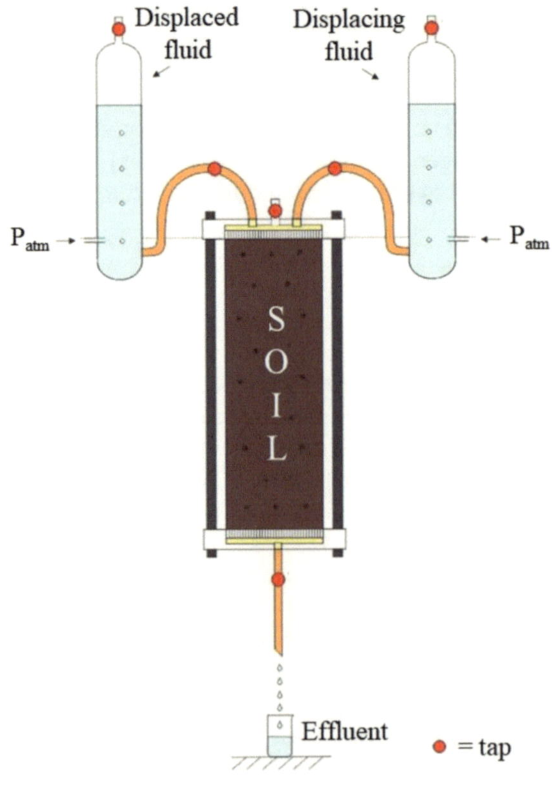

Figure 7.
Schematic representation of the miscible displacement experiment.

Physical-mathematical models are developed to describe solute transport in soil. They are valuable tools in the study of this process, and their success depends on the degree of reliability of the involved variables. The porous matrix controls solute transport through different flow velocities and through different water viscosities in pores of different sizes. The greater the transport distance, the greater the spatial distance of molecules or colloids transported in coarse and fine pores, respectively [57]. The principal equation describing solute transport in soils is the convection-dispersion equation (CDE), in which solute movement is a result of several processes, including diffusion and microscale variations in water flow velocity [58]. Transport models of a chemical species in the soil–air–water system are considered using the CDE [59].

The CDE is nothing more than the continuity equation in which the solute total flux density in the soil (J) is equal to the sum of the solute diffusion flux density (f) and the solute convection flux density (mass flux) (j). Thus, in the vertical direction:

$$\frac{d\theta C}{dt} = -\frac{dJ}{dz} \tag{8}$$

Since

$$J = f + j = -D'\frac{dC}{dz} + qC,$$

then

$$\frac{d\theta C}{dt} = -\frac{d}{dz}\left[-D'\frac{dC}{dz} + qC\right] \tag{9}$$

In this equation:
θ = soil water content (m^3 *of solution*/m^3 *of soil*),
C = solute concentration (*kg of solute*/m^3 *of solution*),
D' = coefficient of diffusion (dispersion) in soil (m^3 *of solution/m of soil.s*),
q = flow density of solution (water) in soil (m^3 *of solution*/m^2 *of soil.s*),
t = time (s), and.
z = vertical coordinate of position (*m of soil*).
Assuming that θ, D', and q are constant, Eq. (9) becomes:

$$\frac{dC}{dt} = D\frac{d^2C}{dz^2} - \bar{v}\frac{dC}{dz} \tag{10}$$

where $D = D'/\theta$ and $\bar{v} = q/\theta$ is the mean solution (water) speed in the soil, because the soil pores are not all the same size.

The differential Eq. (10) is for the model of solute miscible displacement in the soil, where there is neither reaction nor adsorption.

Several solutions and models for fitting data from BTCs are presented by van Genuchten et al. [60] in the software STANMOD (STudio of ANalytical MODels). The STANMOD is an interesting tool to assess different analytical solutions based on CDE for solute transport in porous media. The software integrates seven separate codes with a wide range of solute transport applications. For example, models that assume equilibrium or non-equilibrium transport conditions during steady unidirectional water flow and models that assume non-equilibrium transport.

The greatest interest in solute transport studies in soils has been in the field of agronomy, because the influence of fertilizers, nutrients, and salt concentrations is important for plant growth. However, in recent decades, the interest in solute transport in environment soil science has increased because of the risk of chemical contamination in soil and groundwater caused by agricultural and other chemical substances [59]. Simonin et al. [61] studied the influence of texture and soil organic matter content of six agricultural soils on the transport of two nanoparticles (CuO-NPs and TiO_2-NPs). The solute transport experiments were carried out in water-saturated soil columns, and the breakthrough curve data were fitted to the CDE model with the aim of identifying the dominant process of nanoparticle transfer. The main results show differences between the transport of TiO_2 and CuO. The mobility of TiO_2 was low in all conditions, suggesting a low risk of groundwater contamination. On the other hand, CuO nanoparticles were mobile in all soils. Thus, in the contrasting agricultural soils, the transport of CuO-NPs was mainly controlled by solutes dissolved in the soil solution compared to the solid phase.

Phosphorus (P) and nitrogen (N) are important sources of fertilizers used in agriculture. In some cases, P and N may be associated to risks of potential groundwater contamination, and therefore, indiscriminate use of these sources in agricultural soils should not be recommended. Mahmood-Ul-Hassan [62] assesses the effect of soil structure on the potential for nitrate (NO_3^-) and phosphate ($H_2PO_4^-$) leaching. Laboratory leaching experiments were carried out on undisturbed columns sampled from two types of calcareous soils, Aridisols and Entisols. The leaching curves were fitted with CDE models to determine transport parameters. In general, for Aridisols, the NO_3^- and $H_2PO_4^-$ were found in the effluent immediately after pulse application, whereas for Entisol, NO_3^- was found in the effluent delayed. The experiments showed the importance of knowing the variability of soil structure in predicting the fertilizer losses. Kolachi and Jalali [63] evaluated the movement and retention of P in the columns of two calcareous soils. The concentrations of P and other elements such as K^+, Ca^{2+}, and Mg^{2+} were also determined. P leaching was influenced by supersaturation of P-Ca minerals and undersaturation of Mg-P minerals. BTCs of P and K^+ showed a good fit to equilibrium models using the code CXFIT of STANMOD. The mobility of P in these types of soils may reflect a greater downward movement of water soluble P; that is, these soils require adequate P management to avoid leaching and contamination of groundwater and surface water.

5. Influence of management systems on the movement of water and solutes: effects of the soil compaction

The use of practices, techniques, and treatments to improve soil functionality is known as soil management. Soil management should maintain soil health in favorable conditions to sustain soil functions [64]. The impacts caused by soil use and management can alter and change the hydraulic properties of the soil due to changes in soil structure [65].

The flow of water and solutes is influenced by soil structure, which can vary depending on the type of agricultural management, that is, conventional tillage, no-tillage, crop rotation, cover crops, and others. Soil conservation practices have numerous benefits (e.g., increasing soil resistance to erosive processes, promoting soil quality, and increasing soil aggregation, porosity, water retention, and water infiltration) [66]. These management practices help to maintain long-term soil productivity, which is essential for successful crop production and environmental protection. During the soil compaction processes, the stress applied to the soil results in an increase in the soil bulk density as air is forced out of the pores between the soil particles. In other words, soil pores, particularly macropores, are disrupted and the surface aggregates are degraded, which is referred to as soil compaction [67]. Therefore, machinery traffic can cause disturbances in the soil aeration, water retention, and water conduction by reducing soil pore volume and pore size distribution [68]. Compaction typically occurs when large machinery compacts the soil, but it can also occur when animals walk through the area. As a result, runoff and erosion increase, while the capacity of the affected soil to absorb raindrops decreases.

The use of cover crops and the increasing of SOM as a function of soil management is essential to protect the soil surface from the direct impact of rain drop [69]. The ability of soil to support life and maintain environmental sustainability depends on soil structure [70]. The soil structure is a result of aggregation processes of the soil particles and the arrangement of pore space between aggregates [71]. Soil aggregation is a consequence of several processes mainly influenced by soil inherent attributes (e.g., amounts of sand, silt, and clay particles and soil mineralogy) and dynamic attributes (e.g., amounts of binding agents such as carbonates, soil organic matter, and biological activity). Thus, soil aggregation is a result of chemical, physical, and biological soil interactions.

6. Perspectives for studies related to movement of water and solutes in soils

The use of non-destructive techniques such as X-ray computed tomography (XCT) is expected to advance studies of water and solute movement in soils. 2D and 3D image analysis allows for more detailed quantification of pore space architecture and understanding of soil functions [72]. In particular, studies using 3D image analysis applied to soil science specifically in soil physics have increased in recent decades. The advantages of the XCT are the quantification of parameters such as connectivity, tortuosity, intrinsic permeability, pore morphology, pore area, pore volume, and pore size distribution [73, 74]. Some parameters are important for predicting the movement of water and solutes in soils and cannot be quantified by conventional analyses such as the soil water retention curve (SWRC) and application of capillary theory [75].

Smet et al. [76] investigated how the pore geometry and connectivity of the air-filled pore space change as soils are progressively dried. Twenty samples of a loamy soil were equilibrated at different matric potentials and scanned by X-ray microtomography. The results showed an increase in connectivity and a decrease in tortuosity as the matric potential was lowered. Pessoa et al. [75] studied the behavior of *SWRC* and K_{sat} in highly weathered soils influenced by parameters obtained from 3D image analysis. It was observed a better connectivity for the biggest pore, confirmed by the more negative values of the Euler Number ($EN_{bigpore}$). The better-connected pore space increased the K_{sat} and changed the shape of the SWRCs.

Anderson et al. [77] used XCT to evaluate chemical transport into groundwater through undisturbed geomedia samples. The XCT-based methods presented similar dispersivity values, while the flow method presented high values, a behavior related to additional dispersion promoted by the plate at the end of the column. XCT was efficient in predicting solute transport parameters. Recently, Pak et al. [78] have presented an interesting approach to the technical challenges of performing of XCT-based fluid flow experiments. The authors also discussed the high spatial and temporal resolution that can be achieved with synchrotron imaging sources.

XCT is an important tool to better understand the movement of water and solutes in the soil. However, there are some limitations in the use of the technique, such as strategies to improve image processing (i.e., reconstruction, segmentation, quantification, and simulations), in addition to high computational requirements for processing [79]. Currently, deep learning and machine learning strategies are used for fast and efficient image processing. This type of data analysis allows extracting information from complex datasets [80–82]. Important detailed information on the use of computed tomography applied to soil physics can be found in Mooney et al. [83].

7. Conclusion

Agricultural systems can provide negative impacts on soil properties, but practices such as crop rotation, crop succession, and no-till that improve soil organic matter can reverse negative impacts and maintain good soil physical, chemical, and biological properties. Soil organic matter increases aggregate stability and consequently modifies pore space dynamics. Soil structure controls the movement of water and solutes. The lack of conservation practices in agricultural soil management can lead to soil erosion processes and increase the transport of contaminants to groundwater.

Models for predicting water movement and solute transport in soils are important for understanding different scenarios of groundwater contamination and for improving appropriate soil management. Soil management needs to provide a land use that leads to sustainable agricultural systems, in addition to supporting crop development and food production.

Acknowledgements

The authors would like to thank the Coordenação de Aperfeiçoamento de Pessoal de Nível Superior - CAPES (financing code 001) for supporting the publication of this study.

Conflict of interest

The authors declare that they have no conflicts of interest that could influence the publication of this chapter.

DOI: http://dx.doi.org/10.5772/intechopen.114086

Author details

Thaís Nascimento Pessoa*, Aline Martineli Batista, Rodrigo Nogueira de Sousa and Paulo Leonel Libardi
Department of Soil Science, Luiz de Queiroz College of Agriculture, University of São Paulo, Brazil

*Address all correspondence to: thaisnpessoa@usp.br

References

[1] Ferreira S, Seifollahi AG, Destouni N, Ghajarnia ZK. Soil degradation in the Europeanmediterraneanregion:Processes, status and consequences. Science of the Total Environment. 2022;**805**:1-25

[2] Pfeiffer WC, Lacerda LD, Malm O, Souza CMM, Da Silveira EG, Bastos WR. Mercury concentrations in inland waters of gold-mining areas in Rondônia, Brazil. Science of the Total Environment. 1989;**87-88**:233-240

[3] Mulla DJ, Strock JS, Schepers JS, Raun WR. Nitrogen transport processes in soil. In: Schepers S, Raun WR, editors. Agronomy Monographs: Madison, Wisconsin, United States; 2008. DOI: 10.2134/agronmonogr49.c10

[4] Batista AM, Libardi PL, Giarola NFB. Evaluation of the soil aggregation induced by the plant roots in an oxisol by turbidimetry and water percolation. Rhizosphere. 2020;**16**:100265

[5] Tisdall JM, Oades JM. Organic matter and water-stable aggregates in soils. Journal of Soil Science. 1982;**33**(2):141-163

[6] Moreno-Espíndola IP, Rivera-Becerril F, de Jesús Ferrara-Guerrero M, De León-González F. Role of root-hairs and hyphae in adhesion of sand particles. Soil Biology and Biochemistry. 2007;**39**(10):2520-2526

[7] Carminati A, Zarebanadkouki M, Kroener E, Ahmed MA, Holz M. Biophysicalrhizosphereprocessesaffecting root water uptake. Annals of Botany. 2016;**118**(4):561-571

[8] Van Veelen A, Tourell MC, Koebernick N, Pileio G, Roose T. Correlative visualization of root mucilage degradation using X-ray CT and MRI. Frontiers in Environmental Science. 2018;**6**:32

[9] Chen R, Huang JW, Chen ZK, Xu Y, Liu J, Ge YH. Effect of root density of wheat and okra on hydraulic properties of an unsaturated compacted loam. European Journal of Soil Science. 2019;**70**(3):493-506

[10] Zarebanadkouki M, Ahmed MA, Carminati A. Hydraulic conductivity of the root-soil interface of lupin in sandy soil after drying and rewetting. Plant and Soil. 2016;**398**:267-280

[11] Chang CC, Cheng DH, Qiao XY. Improving estimation of pore size distribution to predict the soil water retention curve from its particle size. Geoderma. 2019;**340**:206-212

[12] Dyck MF. Physics of Water Flow and Solute Transport across a Horizon Interface. Edmonton, AB, Canada: Ph.D. Univ. of Alberta; 2008

[13] Klute A, Dirksen C. Hydraulic conductivity and diffusivity: Laboratory methods. In: Klute A. Methods of Soil Analysis: Part 1 Physical and Mineralogical Methods. Vol. 5. Hoboken, Nova Jersey, United States: Wiley Online Library; 1986. pp. 687- 734. DOI: 10.2136/sssabookser5.1.2ed.c28

[14] Zhang S, Grip H, Lövdahl L. Effect of soil compaction on hydraulic properties of two loess soils in China. Soil and Tillage Research. 2006;**90**(1-2):117-125

[15] Kuncoro PH, Koga K, Satta N, Muto Y. A study on the effect of compaction on transport properties of soil gas and water I: Relative gas diffusivity, air permeability, and saturated hydraulic conductivity. Soil and Tillage Research. 2014;**143**:172-179

[16] Fu Z, Hu W, Beare M, Thomas S, Carrick S, Dando J, et al. Land use effects on soil hydraulic properties and the contribution of soil organic carbon. Journal of Hydrology. 2021;**602**:126741

[17] Bagnall DK, Morgan CL, Bean GM, Liptzin D, Cappellazzi SB, Cope M, et al. Selecting soil hydraulic properties as indicators of soil health: Measurement response to management and site characteristics. Soil Science Society of America Journal. 2022;**86**(5):1206-1226

[18] Simmons CT. Henry Darcy (1803-1858): Immortalised by his scientific legacy. Hydrogeology Journal. 2008;**16**(6):1023-1038

[19] Buckingham E. Studies on the Movement of Soil Moisture. US Dept. Agic. Bur. Soils Bull; 1907. p. 38

[20] Richards LA. The usefulness of capillary potential to soil-moisture and plant investigators. Journal of Agriculture Research. 1928;**37**:719-742

[21] Klute A. Laboratory measurement of hydraulic conductivity of saturated soil. Methods of soil analysis: Part 1 physical and mineralogical properties. Including Statistics of Measurement and Sampling. 1965;**9**:210-221

[22] Deb SK, Shukla MK. Variability of hydraulic conductivity due to multiple factors. American Journal of Environmental Sciences. 2012;**8**(5):489

[23] Dettmann U, Bechtold M, Frahm E, Tiemeyer B. On the applicability of unimodal and bimodal van Genuchten–Mualem based models to peat and other organic soils under evaporation conditions. Journal of Hydrology. 2014;**515**:103-115

[24] Mualem Y. A new model for predicting the hydraulic conductivity of unsaturated porous media. Water Resources Research. 1976;**12**(3):513-522

[25] van Genuchten MT. A closed-form equation for predicting the hydraulic conductivity of unsaturated soils. Soil Science Society of America Journal. 1980;**44**(5):892-898

[26] Assouline S. A model for soil relative hydraulic conductivity based on the water retention characteristic curve. Water Resources Research. 2001;**37**(2):265-271

[27] Tian K, Likos WJ, Benson CH. Polymer elution and hydraulic conductivity of bentonite–polymer composite geosynthetic clay liners. Journal of Geotechnical and Geoenvironmental Engineering. 2019;**145**(10):04019071

[28] Luo Z, Kong J, Shen C, Lu C, Hua G, Zhao Z, et al. Evaluation and application of the modified van Genuchten function for unsaturated porous media. Journal of Hydrology. 2019;**571**:279-287

[29] Ahuja LR. Modeling soluble chemical transfer to runoff with rainfall impact as a diffusion process. Soil Science Society of American Journal. 1990;**54**:312-321

[30] Kelishadi H, Mosaddeghi MR, Hajabbasi MA, Ayoubi S. Near-saturated soil hydraulic properties as influenced by land use management systems in Koohrang region of central Zagros, Iran. Geoderma. 2014;**213**:426-434

[31] Hillel D. Introduction to soil physics. In: Hillel D, editor. Solute Movement and Soil Salinity. Vol. 364. New York: Academic Press; 2003. pp. 167-186

[32] Libardi P. Dinâmica da Água no Solo. São Paulo: Editora da Universidade de São Paulo (EDUSP); 2018

[33] Lal R, Shukla MK. Principles of Soil Physics. Boca Raton, Florida,

United States: Taylor & Francis Group, CRC Press; 2004. DOI: 10.4324/9780203021231

[34] Weil RR, Brady NC. The Nature and Properties of Soils. 15th ed. United Kingdom: Pearson Education; 2017

[35] Costa AD, Albuquerque JA, Costa AD, Pértile P, Silva FRD. Water retention and availability in soils of the state of Santa Catarina-Brazil: Effect of textural classes, soil classes and lithology. Revista Brasileira de Ciência do Solo. 2013;**37**:1535-1548

[36] Lal R. Soil organic matter and water retention. Agronomy Journal. 2020;**112**(5):3265-3277

[37] Moura ED, Moura NG, Marques ES, Pinheiro KM, Costa Sobrinho JRS, Aguiar ACF. Evaluating chemical and physical quality indicators for a structurally fragile tropical soil. Soil Use and Management. 2009;**25**:368-375

[38] Keller T, Da Silva AP, Tormena CA, Giarola NFB, Cavalieri KMV, Stettler M, et al. Soil flex-LLWR: Linking a soil compaction model with the least limiting water range concept. Soil Use and Management. 2015;**31**(2):321-329

[39] de Jong Van Lier Q. Disponibilidade de água às plantas. In: de Jong Van Lier, Q, editors. Física do Solo. 1st ed. Viçosa, Brazil: Sociedade Brasileira de Ciência do Solo; 2010. p. 298

[40] Letey JOHN. Relationship between soil physical properties and crop production. Advances in Soil Sciences. 1985;**1**:277-294

[41] Da Silva AP, Kay BD, Perfect E. Characterization of the least limiting water range of soils. Soil Science Society of America Journal. 1994;**58**(6):1775-1781

[42] Rabot E, Wiesmeier M, Schlüter S, Vogel HJ. Soil structure as an indicator of soil functions: A review. Geoderma. 2018;**314**:122-137

[43] Fernández R, Quiroga A, Lobartini C, Noellemeyer E. Determinación del intervalo hídrico óptimo en usoles de la región semiárida pampeana. Ciencia del suelo. 2017;**35**(2):215-227

[44] Silva BM, Oliveira GC, Serafim ME, Silva EA, Ferreira MM, Norton LD, et al. Critical soil moisture range for a coffee crop in an oxidic latosol as affected by soil management. Soil and Tillage Research. 2015;**189**:36-43. DOI: 10.1016/j.still.2015.06.013

[45] Moura MS, Silva BM, Mota PK, Borghi E, de Resende AV, Acuña-Guzman SF, et al. Soil management and diverse crop rotation can mitigate early-stage no-till compaction and improve least limiting water range in a Ferralsol. Agricultural Water Management. 2021;**243**:106523

[46] Silva BM, Oliveira GC, Serafim ME, Silva ÉA, Guimarães PTG, Melo LBB, et al. Soil moisture associated with least limiting water range, leaf water potential, initial growth and yield of coffee as affected by soil management system. Soil and Tillage Research. 2019;**189**:36-43

[47] Li S, Wu X, Liang G, Gao L, Wang B, Lu J, et al. Is least limiting water range a useful indicator of the impact of tillage management on maize yield? Soil and Tillage Research. 2020;**199**:104602

[48] Oliveira IN, de Souza ZM, Lovera LH, Farhate CVV, Lima EDS, Esteban DAA, et al. Least limiting water range as influenced by tillage and cover crop. Agricultural Water Management. 2019;**225**:105777

[49] Nielsen DR, Biggar JW. Miscible displacement in soils: I. Experimental information. Soil Science Society of America Journal. 1961;**25**(1):1-5

[50] Shaharoona B, Al-Ismaily S, Al-Mayahi A, Al-Harrasi N, Al-Kindi R, Al-Sulaimi A, et al. The role of urbanization in soil and groundwater contamination by heavy metals and pathogenic bacteria: A case study from Oman. Heliyon. 2019;**5**:1771-1782

[51] Pitt RE, Clark S, Parmer K, Field R. Groundwater Contamination from Stormwater Infiltration. Chelsea: Ann Arbor Press, Inc; 2023

[52] Vereecken H, Schnepf A, Hopmans J, Javaux M, Or D, Roose T, et al. Modeling soil processes: Review, key challenges, and new perspectives. Vadose Zone Journal. 2016;**15**(5):1-57

[53] Tuli A, Kosugi K, Hopmans JW. Simultaneous scaling of soil water retention and unsaturated hydraulic conductivity functions assuming lognormal pore-size distribution. Advanced on Water Resources. 2001;**24**:677-688

[54] Nielsen DR, Biggar JW. Miscible displacement: III. Theoretical considerations. Soil Science Society of America Journal. 1962;**26**(3):216-221

[55] Reichardt K, Timm LC. Soil, Plant and Atmosphere. Cham, Switzerland, Springer Cham: Springer Nature Switzerland; 2014

[56] Nielsen DR, Biggar JW. Miscible displacement: IV. Mixing in glass beads. Soil Science Society of America Journal. 1963;**27**(1):10-13

[57] Hartge KH, Horn R, Horton R, Bachmann J, Peth S. Essential Soil Physics. Schweizerbart Science Publishers. Australia: CSIRO Publishing; 2016

[58] Leij FJ, van Genuchten MT. Principles of solute transport. Agricultural Drainage. 1999;**38**:331-359

[59] Skaggs TH, Leij FJ. 6.3 solute transport: Theoretical background. In: Dane JH, Topp GC. Methods of Soil Analysis: Part 4 Physical Methods. Vol. 5. Hoboken, Nova Jersey, United States: Wiley Online Library; 2002. pp .1353-1380. DOI: 10.2136/sssabookser5.4.c56

[60] van Genuchten MT, Šimunek J, Leij FJ, Toride N, Šejna M. STANMOD: Model use, calibration, and validation. Transactions of the ASABE. 2012;**55**(4):1355-1366

[61] Simonin M, Martins JM, Uzu G, Spadini L, Navel A, Richaume A. Low mobility of CuO and TiO_2 nanoparticles in agricultural soils of contrasting texture and organic matter content. Science of the Total Environment. 2021;**783**:146952

[62] Mahmood-Ul-Hassan M, Rashid M, Akhtar MS, Rafique E. Nitrate and phosphate leaching from aridisols and entisols: Laboratory studies and field observations. Soil and Sediment Contamination. 2010;**19**(3):261-276

[63] Kolahchi Z, Jalali M. Phosphorus movement and retention by two calcareous soils. Soil and Sediment Contamination: An International Journal. 2013;**22**(1):21-38

[64] Mitra S, Chakraborty AJ, Tareq AM, Emran TB, Nainu F, Khusro A, et al. Impact of heavy metals on the environment and human health: Novel therapeutic insights to counter the toxicity. Journal of King Saud University-Science. 2022;**34**(3):101865

[65] Chandrasekhar P, Kreiselmeier J, Schwen A, Weninger T, Julich S, Feger KH, et al. Why we should include soil structural dynamics of agricultural

soils in hydrological models. Water. 2018;**10**:1862

[66] Baker CJ, Saxton KE, Ritchie WR, Chamen TWC, Reicosky DC, Ribeiro SMF, et al. No Tillage Seeding in Conservation Agriculture. Wallingford: CABI; 2006

[67] Ampoorter E, de Schrijver A, van Nevel L, Hermy M, Verheyen K. Impact of mechanized harvesting on compaction of sandy and clayey forest soils: Results of a meta-analysis. Annals of Forest Science. 2012;**69**:533-542

[68] Berli M, Kulli B, Attinger W, Keller M, Leuenberger J, Flühler H, et al. Compaction of agricultural and forest soils by tracked heavy construction machinery. Soil and Tillage Research. 2003;**75**:37-52

[69] Bot A, Benites J. The Importance of Soil Organic Matter: Key to Drought-Resistant Soil and Sustained Food Production. Vol. 80. FAO, Rome: Food and Agriculture Organization of the United Nations Soils Bulletin; 2005. pp. 1-80

[70] Bronick C, Lal R. Soil structure and management: A review. Geoderma. 2005;**124**:3-22

[71] Duiker SW, Rhoton FE, Torrent J, Smeck NE, Lal R. Iron (hydr) oxide crystallinity effects on soil aggregation. Soil Science Society of America Journal. 2003;**67**:606-611

[72] Vogel HJ, Balseiro-Romero M, Kravchenko A, Otten W, Pot V, Schlüter S, et al. A holistic perspective on soil architecture is needed as a key to soil functions. European Journal of Soil Science. 2022;**73**(1):e13152

[73] Ferreira TR, Archilha NL, Pires LF. An analysis of three XCT-based methods to determine the intrinsic permeability of soil aggregates. Journal of Hydrology. 2022;**612**:128024

[74] Pessoa TN, Cooper M, Nunes MR, Uteau D, Peth S, Vaz CMP, et al. 2D and 3D techniques to assess the structure and porosity of oxisols and their correlations with other soil properties. Catena. 2022;**210**:105899

[75] Pessoa TN, Ferreira TR, Pires LF, Cooper M, Uteau D, Peth S, et al. X-ray microtomography for investigating pore space and its relation to water retention and conduction in highly weathered soils. Agriculture. 2023;**13**(1):28

[76] Smet S, Plougonven E, Léonard A, Degré A. Characterization by X-ray μCT of the air-filled porosity of an agricultural soil at different matric potentials. European Journal of Soil Science. 2023;**74**:e13387

[77] Anderson SH, Heinze DJ, Peyton RL. Computed tomography-estimated transport velocity and chemical dispersivity in undisturbed geomedia. Procedia Computer Science. 2014;**36**:643-648

[78] Pak T, Archilha NL, Berg S, Butler IB. Design considerations for dynamic fluid flow in porous media experiments using X-ray computed micro tomography–a review. Tomography of Materials and Structures. 2023;**3**:100017

[79] Kumar S, Chakraborty P, Anderson S. X-ray computed tomography for studying solute transport in soils. In: X-Ray Imaging of the Soil Porous Architecture. Cham: Springer International Publishing; 2022. pp. 99-112

[80] Spina TV, Vasconcelos GJ, Gonçalves HM, Libel GC, Pedrini H, Carvalho TJ, et al. Towards real time segmentation of large-scale 4D micro/nanotomography images in the Sirius

synchrotron light source. Microscopy and Microanalysis. 2018;**24**(S2):90-91

[81] Rippner DA, Raja PV, Earles JM, Momayyezi M, Buchko A, Duong FV, et al. A workflow for segmenting soil and plant X-ray computed tomography images with deep learning in Google's Colaboratory. Frontiers in Plant Science. 2022;**13**:893140

[82] Ferreira TR, Cássaro FA, Zhou H, Pires LF. X-ray computed tomography image processing & segmentation: A case study applying machine learning and deep learning-based strategies. In: X-Ray Imaging of the Soil Porous Architecture. Cham: Springer International Publishing; 2022. pp. 57-80

[83] Mooney SJ, Young IM, Heck RJ, Peth S, editors. X-Ray Imaging of the Soil Porous Architecture. Singapore: Springer Nature; 2022

Chapter 7

Examining the Relationship between Crop Net Returns, Risk, and Conservation Practices

Michael Langemeier

Abstract

This study aims to evaluate the effects of different tillage practices and cover crop options on crop net returns, downside risk, soil loss, and greenhouse gas (GHG) emissions in a corn-soybean rotation in central Illinois. The conceptual framework developed encompasses tradeoffs between net returns, downside risk, soil loss, and greenhouse gas emissions. The conservation system had the highest net return per acre. Crop net return differences were smaller between the conservation tillage and no-till systems than they were between the conservation tillage system with no cover crop and with a cover crop. The no-till and cover crop systems also exhibited more downside risk than the conservation system. However, utilizing the no-till system and the cover crop systems was an effective mechanism to reduce soil loss and greenhouse gas emissions.

Keywords: crop net returns, risk, conservation practices, no-till, cover crops

1. Introduction

Conservation practices such as the utilization of a no-till crop system or cover crops can be used to improve soil health, reduce soil loss, and mitigate net greenhouse gas (GHG) emissions. No-till is the practice of refraining from tilling the soil from harvest of the previous crop to the harvest of the current crop. Cover cropping involves planting a crop for seasonal cover. For example, numerous U.S. farms plant annual or cereal ryegrass in the fall to provide cover during the winter months.

No-till and cover crop practices vary widely among farms. Using U.S. Department of Agriculture-Economic Research Service (USDA-ERS) survey estimates, 27 percent of corn in 2016, 40 percent of soybeans in 2012, and 45 percent of wheat in 2017 utilized no-till cropping systems [1]. Though cover cropping is not as common as no-till systems, its use has been growing. In 2017, 5.1 percent of U.S. acres utilized cover crops [2]. Moreover, with the recent interest in carbon farming, a carbon sequestering change made to some farm practices in exchange for receiving payments for carbon credits, the interest in no-till and cover crop practices has increased.

Given their obvious conservation benefits, why have management practices such as no-till and cover crops not been more widely adopted? As with any new system or

IntechOpen

technology, the benefits have to be weighed against the costs. Benefits may include an improvement in soil health, a reduction in soil loss, and mitigation of net greenhouse gas emissions. Costs may include extra costs associated with the practice (e.g., planting a cover crop), potential reductions in crop yields, and potential increase in net return risk (i.e., increased variability or downside risk of net returns resulting from the adoption of a conservation practice), because there are numerous benefits and costs, it is important to examine tradeoffs between key variables such as net returns, risk, soil health, and net greenhouse gas emissions. In other words, it is seldom possible to adopt a conservation practice that has higher net returns and less risk, and that improves soil health or reduces net greenhouse gas emissions.

The objective of this chapter is to examine the relationship between crop net returns, risk, and conservation practices. The conceptual framework developed can be used to explore crop net returns, risk, soil loss, and greenhouse gas emissions for different conservation practices.

2. Previous research

Given the importance of the no-till and cover crop rotations to the conceptual framework developed in this study, the literature review in this section will briefly discuss previous studies that have examined the impact of tillage practices on net returns, soil health, and net greenhouse gas emissions, and the costs and benefits of cover crops. These discussions will be followed by a brief discussion of carbon markets.

Wade et al. [3] summarized the adoption of conservation practices that could be used to reduce the environmental impact of crop production, improve soil health, and reduce net greenhouse gas emissions. Though practices varied widely among crops and regions, approximately 40 percent of the combined acreage of corn, soybean, wheat, and cotton utilized no-till or strip-till practices. Strip-till is a tillage system that utilizes tillage only in narrow strips where seed are planted. In another USDA-ERS study, Claassen et al. [1] noted that no-till and strip-till used in conjunction with conservation rotations and cover crops have a multitude of soil health benefits. The authors noted that practices varied widely among crops and locations. They noted in general, however, that 70 percent of soybean, 65 percent of corn, 67 percent of wheat, and 40 percent of cotton acres used conservation tillage practices, which included no-till, strip-till, and mulch-till. Mulch-till refers to a technique in which organic materials, such as crop residues or mulch, are left on the soil surface and incorporated into the top layer of soil through tillage. The purpose of mulch-till is to improve soil fertility, moisture retention, and weed suppression while reducing soil erosion.

Bergtold and Sailus [4] edited a book that discussed the production, profitability, and stewardship involving conservation tillage systems in the southeast U.S. One of the chapters in the book focused on conservation economics [5]. The authors of this chapter indicated that in order to achieve goals related to maximizing profit and environmental stewardship requires an understanding of tools such as partial budgeting and/or enterprise budgets.

Practices such as the adoption of a no-till crop rotation often involve a learning curve. This fact was confirmed by Cusser et al. [6] who compared the development and stability of a no-till crop rotation with a conventional till system over a 29-year period. The authors indicated that it took over a decade to ascertain the impacts of the no-till system, underscoring the importance of long-run studies.

Precision Conservation Management [7] summarized gross revenue, production costs, soil loss, and greenhouse gas (GHG) emissions for various tillage systems including no-till, strip-till, 1-pass light, 2-pass light, 2-pass moderate, and 2+ passes. The "light" passes utilized low disturbance tillage. Operator and land return were higher for the 1-pass light, 2-pass light, and 2-pass moderate systems than it was for no-till, strip-till, and 2+ passes. In particular, the return for the no-till system was from $19 to $30 lower than the return for the light and moderate pass systems. However, soil loss and GHG emissions were lower for the no-till system than they were for other tillage systems examined.

Next, we will turn our attention to previous research pertaining to cover crops. Plastina et al. [8] used partial budgets to assess the economic return to cover crop use on Midwest U.S. row crop farms. Essentially, the authors compared returns and costs for farms that utilized cover crops to those that did not utilize cover crops. The economic loss accruing to use of cover crops terminated with herbicides in a corn/soybean rotation was $12 per acre with the inclusion of cost-share payments and $43 per acre without the inclusion of these payments. Plastina et al. [9] used focus groups and a partial budgeting framework to compare the net returns of crop systems with and without the use of cover crops. The net change in profits associated with cover crop use was a negative $54 per acre.

Myers et al. [10] note that additional net returns generated from the adoption of cover crops is difficult to discern with one-year budget analysis. The authors noted that some farms have a very long-term positive experience associated with the use of cover crops. Cover crops were found to provide a faster payment when herbicide-resistant weeds are a problem, cover crops are grazed, soil compaction is an issue, cover crops are used to speed up and ease the transition to no-till, soil moisture is at a deficit, fertilizer costs are high, and incentive payments are available.

A recent USDA-ERS publication noted that the share of harvested cropland with cover crops increased from 3.4 percent in 2012 to 5.1 percent in 2017 [2]. This study noted that unlike no-till and conservation tillage, cover crops involve increased costs, particularly in the short term. A recent survey by the Conservation Tillage Information Center [11] indicated that cover crops are typically combined with a no-till system. Approximately 12 percent of those surveyed went from not using cover crops to using cover crops on some of their acres between 2015 and 2019. Over half of the respondents used cover crops on at least 40 percent of their crop acres.

Gross revenue, production costs, soil loss, and GHG emissions for a no cover crop system and two cover crop systems, overwintering and winter terminal, are summarized by Precision Conservation Management [7] and Sellars et al. [12]. Operator and land return for the overwintering and winter terminal cover crop systems used for corn production (soybean production) were $32 ($44) and $12 per acre ($21 per acre) lower than the return under the no cover crop system, respectively. As with the utilization of a no-till system, soil loss and GHG emissions were reduced by utilizing the cover crop systems.

The potential to sequester carbon in agricultural soils has increased public and private interest in markets that pay farmers to sequester carbon using various management practices. Two common management practices discussed are the adoption of a no-till system or the use of cover crops. In their review of the literature, Havens et al. [13] noted that the average implementation cost for no-till was $17 per acre and that the average implementation cost for cover crops was $31 per acre. On average, sequestration resulted from the adoption of a no-till system and cover crops would be 0.77 and 0.76 metric tons per acre. In addition to discussing the average sequestration and average cost per acre, the authors estimated that the social benefit of sequestration would be $51 per metric ton of CO_2, which would more than offset the expected cost of implementing no-till or cover crop systems.

Thompson et al. [14] reported recent survey results pertaining to farm adoption of carbon contracts. Of the survey respondents, 39 percent were aware of opportunities to receive payments for capturing carbon, 7.1 percent had actively engaged in discussions regarding receiving payments for capturing carbon, and 1.3 percent had signed a contract to capture carbon. In a follow-up question, reasons preventing producers from enrolling, in order of percentage responding affirmatively, were payment level offered that was not large enough, potential legal liability, skepticism of carbon sequestration viability, and previous use of eligible practices were not covered under the contract.

In a series of Ag Decision Maker articles, Plastina [15], Plastina and Jo [16], and Plastina and Wongpiyabovorn [17] discuss carbon markets; how data, payments, and methods are made in carbon programs; and describe a tool that can be used to examine the potential net returns for carbon farming. Net returns are dependent on farming practices, contract type, contract length, frequency of additional practice implementation, contract price for each practice, expected change in contract price, farm area enrolled, participation in cost share programs, expected changes in costs, and discount rate. Plastina [15] noted that programs typically require additionality to generate payments. Additionality means that farmers must do something different or adopt a new practice to receive carbon payments.

3. Methods

Experimental data that contain all of the variables that need to be explored when examining tradeoffs between key variables are lacking. However, it is possible to develop a conceptual framework and provide an illustration using available data sources pertaining to crop budgets, soil health parameters, and net greenhouse gas emissions.

Crop budgets for 2015–2021 were developed using several sources of information. Crop yield data were obtained from Precision Conservation Management [7], Schnitkey and Swanson [18, 19], and Sellars et al. [12]. Most of the tillage scenarios reported in Precision Conservation Management [7] could be considered conservation tillage systems, so the crop system used to compare with the no-till and cover cropping systems will be referred to as a conservation tillage system. Crop yield data for the conservation tillage system were obtained from Precision Conservation Management [7] and Schnitkey and Swanson [18, 19]. Yields for corn and soybeans produced after a cover crop were obtained from Sellars et al. [12]. Annual yield data were not available for the no-till practice from the sources noted above so additional assumptions were needed to estimate crop yields. For corn, the average difference between corn yields for a no-till and a 1-pass light (i.e., one pass with low disturbance) was used to estimate annual yields. For soybeans, the relationship between soybean crop yields generated under a no-till system and under miscellaneous tillage systems reported by the Center for Farm Financial Management [20] was used to estimate annual yields. Crop price, government payments, and crop insurance indemnity payments were obtained from Schnitkey and Swanson [18, 19].

Cost information for direct costs (fertilizers, pesticides, seed, cover crop seed, drying, storage, and crop insurance), power costs (tillage, fertilizer application, spraying, planting, harvesting, and grain hauling), and overhead costs (hired labor, building repairs and depreciation, insurance, and interest) was obtained from Precision Conservation Management [7] and Schnitkey and Swanson [18, 19]. To account for the use of soybean yields from the Center for Farm Financial Management [20] for the no-till system, the relationship between soybean costs using a no-till system and miscellaneous tillage systems was utilized to estimate soybean costs.

Net return was computed by subtracting direct costs, power costs, and overhead costs from gross revenue, which consisted of crop revenue (crop price × crop yield), government payments, and crop insurance indemnity payments. Net return excludes operator and land costs so it can be interpreted as a net return to operator and land.

For risk averse decision makers, the choice between conservation practices depends on not only the expected net return and soil health and net greenhouse gas emissions, but also the variability in net returns. For example, if a particular conservation practice reduces or increases crop yield variability, this fact will be important to decision makers. One of the ways to incorporate net return variability into the analysis is to compute the certainty equivalent of net returns for each conservation practice. The certainty equivalent of net returns represents a risk-adjusted net return. In other words, it incorporates both expected net return and risk. Certainty equivalents are computed using an expected utility function and specific levels of risk aversion [21]. The power utility function was used to compute certainty equivalents in this study. This utility function has been used extensively to model risk aversion in production agriculture [22]. Relative risk aversion levels of 0, 1, 3, and 5 were used in this study. A relative risk aversion level of 0 is applicable to a risk-neutral decision maker. Someone with these preferences would choose the conservation practice with the highest average net return. Risk aversion levels of 1, 3, and 5 represent slightly, moderately, and strongly risk averse preferences [21].

In addition to comparing the expected net return and certainty equivalents among the conservation practices, a conceptual framework was developed to examine the tradeoffs between net return, risk, soil loss, and GHG emissions. This conceptual framework is consistent with the notion that farms use a multidimensional framework when making decisions [23, 24]. The specific model used represents a modified version of the Target MOTAD model. The Target MOTAD model maximizes expected net return subject to a constraint on downside risk, which is measured as the total negative deviations below a specified target [25, 26]. A target income or net return to land and operator labor of $280 was used. This target income represents the average cash rent over the study period. Constraints related to soil loss and gas emissions were added to the Target MOTAD model so that tradeoffs could be examined. A similar approach was used by Stucky [27] and Stucky et al. [28] to examine the tradeoff between return to labor and management, risk, and water quality for crop rotations in Kansas.

Solutions to the modified Target MOTAD model are generated by relaxing the constraints on downside risk, soil loss, and net greenhouse gas emissions. Each solution contains information on the expected net return, downside risk or the total negative deviations below a specified target, soil loss, and net greenhouse gas emissions.

4. Results and discussion

Table 1 presents the annual net return, average net return, level of downside risk, soil loss, and greenhouse gas (GHG) emissions for each crop and conservation practice. The base case, as noted above, assumes the use of a conservation tillage system.

	Base case		No-till		Cover crop	
Item	**Corn**	**Soybeans**	**Corn**	**Soybeans**	**Corn**	**Soybeans**
2015	232	309	215	206	229	262
2016	233	349	215	316	206	301
2017	212	314	194	309	177	266
2018	297	448	278	455	242	401
2019	287	325	270	340	233	278
2020	425	395	404	377	373	345
2021	696	585	665	667	613	525
Avg NR ($/acre)	340	389	320	381	296	340
Downside risk ($/acre)	−23	0	−32	−11	−45	−5
Soil loss (tons/acre)	0.93	1.29	0.55	0.90	0.64	0.96
GHG emissions (mt C/acre)	0.30	−0.28	−0.32	−1.12	−0.72	−1.76

Table 1.
Crop net return, soil loss, and GHG emissions for alternative conservation practices.

Downside risk in **Table 1** is represented using average annual negative deviations rather than total negative deviations below the target net return of $280. In general, soybeans were more profitable than corn during the study period.

Without considering soil health and net greenhouse gas emissions which will be addressed below, a risk neutral decision maker would choose the conservation practice with the highest net return. Assuming the adoption of a corn/soybean rotation, the net return for corn and soybeans is the highest for the base case. The average difference in net returns between the base case and the no-till system was $20 per acre for corn and $8 per acre for soybeans. The difference between net returns was larger for the comparison between the base case and the cover crop system. For corn, net return using cover crops was $44 per acre lower than the base case. The net return difference for soybeans was even greater at $49 per acre.

Soil loss and net greenhouse gas emissions can be reduced by adopting a no-till or cover crop system. The no-till system results in the largest drop in soil loss, while the cover crop system exhibits the largest reduction in net greenhouse gas emissions.

The certainty equivalent of net return (CE) for each crop and conservation practice is depicted in **Table 2**. The crop enterprise with the #1 label refers to the base case. The crop enterprise labels designated as #2 and #3 refer to the no-till system and the cover crop system, respectively. The differences in CEs among risk aversion levels for corn are similar to the difference in average net returns illustrated in **Table 1**. In contrast, the difference in CEs between soybean enterprises depends on the level of risk aversion. What explains these results? If there is not much difference in annual net returns between conservation practices, which is the case for corn, the CE results will not differ very much from the average net return results. For soybeans, the annual net returns across conservation practices differ more than it does for corn; thus, comparisons between soybean enterprises will depend more on the level of risk aversion. Too see this, let us focus on the comparison between soybean net returns for the base case and soybean net returns for the no-till system. The largest divergence in net returns was for 2015 and 2021. In 2015, soybean net returns were substantially lower for the no-till system. In contrast, for 2021, net returns were relatively lower for the base case.

	Relative risk aversion		
Item	**r = 1**	**r = 3**	**r = 5**
C #1	312	278	261
S #1	380	365	355
C #2	292	259	242
S #2	360	324	295
C #3	271	242	227
S #3	330	314	314
C #1 minus C #2	20	20	19
C #1 minus C #3	41	36	34
S #1 minus S #2	20	41	59
S #1 minus S #3	50	51	41

Table 2.
Certainty equivalent of net returns for corn and soybean enterprises ($/acre).

In summary, results in **Tables 1** and **2** indicate that net returns and risk-adjusted net returns were higher for the base case than they were for the no-till and cover crop systems. Assuming a corn/soybean rotation, the average difference in CEs between the base case and the no-till system was $26, and the average difference in CEs between the base case and the cover crop system was $43. These results did not account for differences in soil loss and GHG emissions. The modified Target MOTAD results discussed below will incorporate differences in these items.

Three sets of results for the modified Target MOTAD model are illustrated in **Table 3**. Scenario #1 represents the maximum expected net return solution. Because soil loss and GHG emissions were not constrained for the base case, the values for these two conservation dimensions are the highest for the base case scenario. Also, note that corn and soybeans produced under the base case are part of a conservation tillage system. The other two scenarios illustrate solutions for a

Item	**Scenario #1**	**Scenario #2**	**Scenario #3**
C #1	0.500	0.151	0.000
S #1	0.500	0.151	0.000
C #2	0.000	0.349	0.231
S #2	0.000	0.349	0.231
C #3	0.000	0.000	0.269
S #3	0.000	0.000	0.269
Expected net return ($/acre)	364.79	355.00	333.09
Negative deviations ($/acre/year)	3.79	11.90	16.85
Soil loss (tons/acre)	1.11	0.84	0.77
GHG emissions (mt C/acre)	0.01	−0.50	−1.00

Table 3.
Expected net return, risk, soil loss, and GHG emissions for alternative scenarios.

GHG emission constraint of −0.50 and −1.00, respectively. As in **Table 2**, the crop enterprise with the #1 label refers to the base case, the crop enterprises with the #2 label refers to the no-till system, and the crop enterprise with the #3 label refers to cover cropping system.

As expected, the maximum net return solution produces corn and soybeans using the conservation tillage system, or the base case. Expected net return for the solution is $365 and average annual negative deviations below the $280 target net return is $4. Soil loss for this solution is 1.11 tons per acre and GHG emissions are 0.01 mt C per acre. With scenario #2, GHG emissions are reduced to −0.50 mt C per acre and soil loss is reduced approximately 24 percent. To achieve these reductions, net return per acre was reduced $10 per acre and downside risk increased from $4 per acre to $12 per acre. Under this scenario, approximately 30 percent of crops were produced using the conservation tillage rotation. The other 70 percent of crops were produced using the no-till system.

With scenario #3, GHG emissions are reduced to −1.00 mt C per acre and soil loss is reduced approximately 31 percent. These reductions were achieved by producing 46 percent of the crops under the no-till system and 54 percent of the crops under the cover cropping system. Not surprisingly, given the difficulty of reducing GHG emissions from 0.01 to −1.00, the reduction in expected net returns and increase in downside risk when comparing the shift from scenario #1 to #2 with the shift from scenario #1 to #3 were larger in this instance. Specifically, net return was reduced by $32 per acre and downside risk increased from $4 per acre to $17 per acre under scenario #3.

There are several extensions that could be made to the conceptual framework presented in this study. First, it would be helpful to include more years in the comparisons among crop systems. This would ensure that we are properly capturing the difference in crop yields and costs between the systems. Given the learning curves associated with adopting a no-till or cover crop rotation, this point is particularly salient [6, 10]. Second, it would be helpful if annual data pertaining to soil loss and GHG emissions were available. With these data, it would be possible to generate another set of constraints that would capture the risk associated with not achieving a target level of soil loss or GHG emissions. With annual data, years with higher soil loss or GHG emissions, due to whatever cause (e.g., weather fluctuations), than a target, which could be represented by the averages for a sample period, would be penalized. For example, using a similar conceptual framework, Stucky et al. [28] examined improvements in water quality one variable at a time. Solutions examining each water quality variable were constrained so that deviations below the target could not increase above those from the net return maximizing solution and by not allowing any of the water quality variables except for the variable of interest to exceed average levels.

5. Conclusions

This chapter examined the relationship between crop net returns and conservation practices. The conceptual framework developed can be used to compare crop net returns, risk, soil loss, and GHG emissions for different conservation practices. Using an example that compared net returns, downside risk, soil loss, and GHG emissions, it was possible to reduce both soil loss and GHG emissions. However, these reductions came at a cost. To achieve a reduction in GHG emissions of approximately −0.50 mt C per acre, it was necessary to produce part of the crops under a no-till system. For this case, expected net return was reduced approximately $10 per acre and downside risk increased approximately $8 per acre compared to the maximum expected net return solution, which involved producing corn and soybeans utilizing a conservation tillage rotation. A reduction in GHG emissions of −1.00 mt C per acre was achieved by utilizing a combination of the no-till and cover crop systems. The reduction in expected net return was approximately $32 per acre, and the increase in downside risk was approximately $13 per acre in this case.

Conflict of interest

The author declares no conflict of interest.

Author details

Michael Langemeier
Purdue University, West Lafayette, Indiana, United States

*Address all correspondence to: mlangeme@purdue.edu

References

[1] Claassen R, Bowman M, McFadden J, Smith D, Wallander S. Tillage Intensity and Conservation Cropping in the United States. Washington, DC: USDA-ERS, Economic Information Bulletin Number 197; 2018

[2] Wallander S, Smith D, Bowman M, Claassen R. Cover Crop Trends, Programs, and Practices in the United States. Washington, DC: USDA-ERS, Economic Information Bulletin Number 222; 2021

[3] Wade T, Claassen R, Wallander S. Conservation Practice Adoption Rates Vary Widely by Crop and Region. Washington, DC: USDA-ERS, Economic Information Bulletin Number 147; 2015

[4] Bergtold J, Sailus M, editors. Conservation Tillage Systems in the Southeast: Production, Profitability, and Stewardship, SARE Handbook Series 152020. Brentwood, MD: SARE (Sustainable Agriculture Research and Extension). pp. 204-218

[5] Bergtold J, Smith A, Lamb M, Duzy L. Conservation economics: Budgeting, cover crops, and government programs. In: Bergtold J, Sailus M, editors. Conservation Tillage Systems in the Southeast. Brentwood, MD: SARE (Sustainable Agriculture Research and Extension); 2020. p. 308

[6] Cusser S, Bahlai C, Swinton SM, Robertson GP, Haddad NM. Long-term research avoids spurious and misleading trends in sustainability attributes of no-till. Global Change Biology. 2020;**26**:3715-3725. DOI: 10.1111/gcb.15080

[7] Precision Conservation Management. The Business Case for Conservation: Cost-Benefit Analysis of Conservation Practices. 2015-2021 Data Summary. Champaign, IL: Precision Conservation Management; 2022

[8] Plastina A, Liu F, Sawadgo W, Miguez FE, Carlson S. Partial budgets for cover crops in Midwest row crop farming. Journal of the American Society of Farm Managers and Rural Appraisers. 2018:90-106

[9] Plastina A, Liu F, Miguez F, Carlson S. Cover crops use in Midwestern U.S. agriculture: Perceived benefits and net returns. Renewable Agriculture and Food Systems. 2018;**35**:38-48. DOI: 10.1017/S1742170518000194

[10] Myers R, Weber A, Tellatin S. Cover Crop Economics: Opportunities to Improve Your Bottom Line in Row Crops. Brentwood, MD: SARE Technical Bulletin; 2019

[11] Conservation Technology Information Center. Annual Report 2019-2020, National Cover Crop Survey. 2020

[12] Sellars SC, Schnitkey GD, Gentry LF. Cover crops on Illinois farms. Journal of the American Society of Farm Managers and Rural Appraisers. 2023 [Forthcoming]

[13] Havens D, Perrin RK, Fulginiti LE. Carbon Farming: A Preliminary Economic Analysis of Carbon Credits for No-till and Cover Crops. Lincoln, NE: Cornhusker Economics, University of Nebraska-Lincoln; 2023

[14] Thompson N, Hughes MN, Nuworsu EK, Reeling CJ, Armstrong SD, Mintert JR, et al. Opportunities and

challenges associated with carbon farming for U.S. row crop production. Choices. 2022;**37**:1-10. DOI: 10.22004/ag.econ.329528

[15] Plastina A. How Do Data and Payments Flow through Ag Carbon Programs. Ames, IA: Ag Decision Maker, Iowa State University, A1-77; 2022

[16] Plastina A, Jo H. Net Returns to Carbon Farming. Ames, IA: Ag Decision Maker, Iowa State University, A1-78; 2023

[17] Plastina A, Wongpiyabovorn O. How to Grow and Sell Carbon Credits in U.S. Agriculture. Ames, IA: Ag Decision Maker, A1-76; 2023

[18] Schnitkey G, Swanson K. Revenue and Costs for Illinois Grain Crops, Actual 2015 through 2021 and Projected 2022. Champaign, IL: Department of Agricultural and Consumer Economics, University of Illinois; 2022

[19] Schnitkey G, Swanson K. Revenue and Costs for Illinois Grain Crops, Actual 2016 through 2021 and Projected 2022 and 2023. Champaign, IL: Department of Agricultural and Consumer Economics, University of Illinois; 2022

[20] Center for Farm Financial Management, University of Minnesota. FINBIN Database. 2023. Available from: https://finbin.umn.edu [Accessed: April 24, 2023]

[21] Hardaker JB, Lien G, Anderson JR, Huirne RBM. Coping with Risk in Agriculture. 3rd ed. Cambridge: CABI Publishing; 2015. p. 276. DOI: 10.1079/9780851998312.0000

[22] Langemeier M, Liu Y. Comparing net return to land and risk for alternative leasing arrangements. Journal of the American Society of Farm Managers and Rural Appraisers. 2020:129-136

[23] Patrick G, Blake BF, Whitaker SH. Farmers' goals: Uni- or multi-dimensional? American Journal of Agricultural Economics. 1983;**65**:315-320. DOI: 10.2307/1240881

[24] Sintori A, Rozakis S, Tsiboukas K. Multiple goals in farmers' decision making. In: Paper Presented at the Annual Conference of the Agricultural Economics Society. Dublin, Ireland; 2009

[25] Tauer L. Target MOTAD. American Journal of Agricultural Economics. 1983;**65**:606-610. DOI: 10.2307/1240515

[26] Watts MJ, Held LJ, Helmers GA. A comparison of target MOTAD to MOTAD. Canadian Journal of Agricultural Economics. 1984;**32**:175-186. DOI: 10.1111/j.1744-7976.1984.tb02009.x

[27] Stucky T. Tradeoff analysis between water quality and net returns in Central and Eastern Kansas [Unpublished M.S. Thesis]. Manhattan, KS: Department of Agricultural Economics, Kansas State University; 2000

[28] Stucky T, Langemeier M, Nelson R, Fjell D, Devlin D. An examination of the tradeoff between return to land and management, risk, and water quality for crop rotations in Northeast Kansas. In: Paper Presented at the Western Agricultural Economics Association Meeting. Rapid City, SD; 2000

Chapter 8

Strategic Tillage for Sustaining the Productivity of Broadacre Cropping in the Arid and Semi-Arid Regions of Southern Australia

Gaus Azam, Md. Shahinur Rahman and Kanch Wickramarachchi

Abstract

Conventional tillage, usually practised before every cropping cycle, was proven damaging and unsustainable and was replaced by conservation agriculture (CA) using no-till systems following the 'dustbowls' incident in the USA. However, the continuous practice of CA has brought new soil and agronomic challenges, such as soil water repellence, soil pH and nutrient stratification, subsoil acidity, compaction and herbicide resistance, threatening the sustainability of broadacre cropping again. In recent years, one-off deep strategic tillage (ST) has brought attention and shown promise in overcoming the challenges imposed by CA and improving the sustainability of broadacre cropping. Deep ST approaches are now available for applying and incorporating soil amendments such as agricultural lime to a targeted depth while treating soil water repellence and loosening the compacted subsoil. Some ST practices have also been proven to manage weed seed banks and decrease the demand for herbicide applications. Many farmers in southern Australia have adopted ST to address the above-mentioned soil and agronomic challenges. When ST is practised, care should be taken in selecting the right timing, soil conditions and depth of tillage for successful outcomes. Once ST is implemented, reestablishing CA would ensure the longevity of the benefits of ST.

Keywords: strategic deep tillage, water-repellent surface soil, subsurface soil acidity, soil compaction, soil penetrometer resistance, nutrient redistribution, soil re-engineering

1. Introduction

Conventional tillage (CT, usually practised in every cropping cycle) in agriculture involves intentional soil manipulation using mechanical means for increasing water infiltration and storage of soil moisture to improve seed germination and root growth, suppressing weed population and mixing crop residues and organic materials.

However, the disadvantages of CT practices include the breaking of soil structure, which might lead to an increase in soil dispersion [1], wind and water erosion [2, 3], loss of conserved soil moisture [3, 4] and reducing soil organic carbon content [5] depending on the depth of tillage.

To reduce soil erosion, moisture loss, preserve soil organic matter, promote good soil structure and better nutrient cycling and plant nutrition, the Food and Agricultural Organisation (FAO) proposed three principles of conservation agricultural (CA) or no-tillage (NT) system. They are: (i) practising minimum soil disturbance for seed and fertiliser placement, (ii) permanently covering at least 30% of the soil surface with organic matter, i.e., crop residue or cover crop, and (iii) diversifying crop species [6]. CA is not a new concept, and it was first conceptualised to protect the soil by the farming community and scientists in the 1930s after the 'dustbowls' incident during a drought in the mid-west of the USA due to extensive cultivation. However, the CA only became more popular with the development of the improved seeding machinery in the late 1940s [5] and the widespread use of herbicides in 1960s in USA [7]. Currently, 12.5% of agricultural lands under the CA practices around the globe, and Australia has adopted CA practices at a wider scale than any other country [8–10]. The Australian grain grower survey in 2016 reported that around 80–90% of the strategically tilled soil does not receive any pre-sowing cultivation [11].

In current agricultural practices, many agronomic and soil constraints such as water-repellent surface soils [12, 13], soil pH and lime stratification have arisen [14], increase in subsurface soil acidity [15–17], nutrient stratification [18], some soil-borne pests and diseases such as slugs [19] or nematodes [20], and herbicide-resistant weeds [21, 22] due to the long-term NT practices. To manage these soil and agronomic constraints, occasional tillage (known as strategic tillage, ST) might be one of the ways, but there is a fear of reversing the benefits of long-term NT practices on soil physical, chemical and biological properties. While the benefits of CA or NT are overarching, the complete elimination of tillage from the agricultural system and its effect on sustaining the productivity of the broadacre cropping system still need to be investigated [23–33]. This book chapter postulates that ST would remain a key element for sustaining the productivity of broadacre cropping, especially in the arid and semi-arid regions of southern Australia.

2. Strategic tillage: pros and cons

To avoid the conventional tillage (CT) practice, a needs-based occasional, usually every 3–10 years, deep tillage approach [34, 35] is gaining popularity, known as strategic tillage (ST). The occasional soil disturbance in a conservation agriculture (CA) or no-till (NT) system could minimise the risk of CT and amend the soil and agronomic constraints. Besides, ST can increase crop yield significantly over a period which might subsidise the cost of tillage operations and make the cropping system more sustainable and profitable than an NT system. However, there is fear that the practice of ST in a CA or NT system might affect the benefit of the long-term CA by affecting soil properties such as soil erosion and runoff, wind erosion, loss of soil aggregate, infiltration of soil water and soil organic carbon.

The impact of ST on soil properties, in a CA or NT system, needs to be more consistent, and still, there are opportunities to conduct more research and explore further. For instance, soil erosion and runoff may be accelerated by ST in an NT system. Usually, soil erosion and runoff depend on soil hydraulic conductivity and other water infiltration properties [36]. Occasional tillage with a Mouldboard plough did not show any difference in soil hydraulic conductivity in a 35-year NT practice system [37]. Usitalo et al. [38] reported that the loss of dissolved phosphorus by runoff was 67% less in a CT system than in an NT system due to reduced runoff and improve infiltration in tilled soil. A recent review [36] reported that ST increased runoff in two out of five studies while decreased runoff in two studies and had no effect in one. Therefore, the ST in an NT system has a mixed effect on soil erosion and runoff and might also depend on other soil properties, such as soil texture and structure.

NT is very effective in wind-prone and semiarid areas in decreasing wind erosion by crop residue [39, 40]. The tillage buried the crop residue and enhanced wind erosion by increasing the emission of particulate matter in the soil (<2.5 to 10 um) [41, 42]. Considering dry soil aggregate stability, an important factor for soil erosion, no immediate effect of tillage was observed after 1 to 3 years on a 10-year NT system in loam, silt clay loam and clay loam soils [27].

The ST might impact soil aggregation, infiltration and soil water content. A review by Blanco-Canqui and Wortmann [36] reported that ST did not affect the wet and dry stability of soil aggregates in two out of three studies and decreased in one study. It has been reported that soil aggregates might decrease immediately after tillage but would reaggregate soon, within 7 to 12 months, if no further soil disturbance occurs immediately [43].

The impact of ST on soil hydraulic conductivity, soil infiltration and water retention are also mixed. ST might not have any effect [44], inconsistent effect [3, 45], or decrease [46], or increase [47] the soil water infiltration rate. It has been reported that frequent tillage impact more negatively than ST in long-term NT soil [48]. ST reduces soil cover, and it is often thought to decrease plant available water in soil through enhanced evaporation due to exposure to the sun and an increase in soil temperature [4]. It has been reported that soil temperature does not vary significantly between NT and ST [27]. Soil water content might decrease immediately after tillage [3], but it would recover quickly after rain. Therefore, the timing of the ST needed to be considered to minimise water loss.

The tillage operation usually impacts soil organic carbon (OC), where an NT will build soil OC on the surface [49]. ST would assist in removing OC stratification from the topsoil and a uniform distribution along the soil profile [14, 50]. Quincke et al. [51] reported an increase in soil OC in 0.10–0.30 m following an event of mouldboard plough in an NT soil. One of the major concerns is that the tillage might break soil aggregates and expose the protected soil OC to microbial decomposition. The effect of ST on soil OC on top 0.10 m was reviewed in 11 studies with 28 soils; out of which 22 soils showed no effect and 6 soils decreased the level of OC [36]. These results might indicate a limited effect of ST on soil OC content. However, even if soil OC decreases near the surface it has been reported that OC would build up below the surface due to the inversion or mixing of topsoil containing high OC with the subsoil that has low OC [14, 30, 46, 47, 51] as well as through enhanced plant root growth. Where a decrease in OC is reported due to the use of ST, an increase in soil nutrient status was also reported which attributed to the mineralisation of nutrients from the incorporated soil OC and improving crop growth and yield following an event of ST [36].

Another concern of the ST is its effect on the soil microbiological community. The review by Blanco-Canqui and Wortmann [36] reported a small or no effect of ST on soil microbial community in four studies. Dang et al. [3] reported no effect of ST within a few months of using a chisel plough on soil microbial activity or biomass in Australia. In contrast, Wortmann et al. [30, 52] reported a persistent reduction in soil microbial activity in 5 years in Nebraska, USA. However, the decrease in soil microbial activity did not affect the crop yield which might indicate that ST did not have any effect or minimal effect on the broader soil ecosystem [30]. Furthermore, Garcia et al. [53] reported that ST reduced root colonisation by arbuscular mycorrhizae but did not decrease the phosphorus uptake by plant roots.

3. Role of strategic tillage in southern Australian agriculture

In this section we focus on the role of ST in removing soil and agronomic constraints and providing additional benefits toward maintaining sustainable productivity in a broadacre cropping system. A cause-and-effect relationship between multiple interacting soil constraints (soil water repellence, compaction, acidity and sodicity) arises due to long-term farming practices with NT adoption by farmers and the reasons for generating a yield gap are presented in **Figure 1**. The role of ST in amelioration of these constraints such as water repellence, soil acidity and compaction will be discussed further in this chapter.

3.1 Role of ST for amelioration of soil and agronomic constraints

3.1.1 ST for treating water-repellent surface soils

Water repellence is one of the major soil constraints in southern Australia. It is estimated that around 10.2 Mha of arable land in the south-west of Western Australia is at risk of water repellence with 3.3 Mha marked as at high risk and another 6.9 Mha at moderate risk [58]. The main reason for water repellence is due to an increased OC content (hydrophobic carbon compounds such as plant waxes and other products from natural processes of plant biodegradation) in the surface soil [59–62], especially

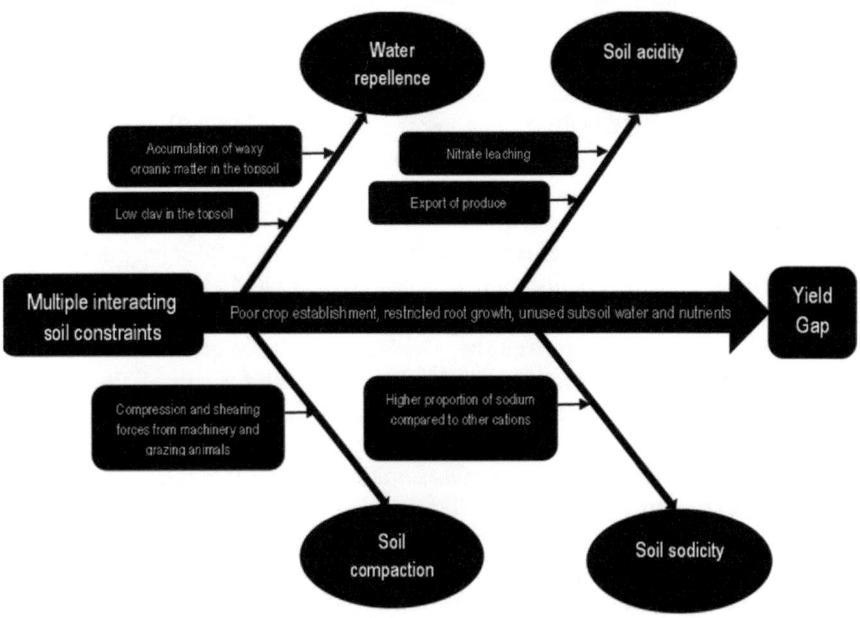

Figure 1.
Cause and effect relationship of multiple interacting soil constraints. For further information on soil constraints' cause and effects, see soil quality: Ebook series [54–57].

in soils with low clay content, i.e., soil with a low particle surface area [60, 63]. The CA or NT practice has exacerbated the scenario by concentrating organic matter and its wax component in the top 0.05 m of the soil [55].

A set of agronomic practices such as furrow or on-row seeding, and repeated application of wetting agents can temporarily treat soil water repellence. Using ST such as mouldboard ploughing, spading, etc. can be a medium-to-long-term solution for treating soil water repellence, albeit at a moderate to high cost. As soil water repellence mainly occurs in sandy topsoil with less than 5% clay [55], the addition of clay followed by incorporation using an ST is a longer-term solution. Claying can be expensive; however, this cost can be cheaper if the clay source is within the same paddock, especially where texture contrast soils (e.g. Planosol and Solonetz in the World Reference Base for Soil Resources) are available. When clay is applied to treat soil water repellence, it is essential to ensure the total amount of clay (applied plus original clay in soil matrix) is around $7 \pm 2\%$. Any less than this amount could treat soil water repellence inefficiently. In contrast, a higher amount might enhance surface crusting and decrease plant available water in the surface soil, especially during plant emergence [55].

One-off deep cultivation, ST, aimed to mix topsoil with subsoil, was found effective in decreasing soil water repellence [64]. However, if the surface soil is extremely repellent, i.e., soil with a molarity of ethanol droplet (MED) value of >3 [55], there is a risk of making whole soil profile water repellence due to soil mixing as proposed by Steenhuis et al. [65]. Mouldboard ploughing has been reported to effectively decrease soil water repellence by burying the repellent topsoil and bringing wettable subsoil to the surface [66, 67]. Davies et al. [68] found rotary spading also effectively decreased soil water repellence in 12 field trial sites in Western Australia. A study by Davies et al. [69] compared different ST implements for mixing soil and treating water repellence and found that benefits from different implements varied in the order of offset disc < deep ripped < rotary hoe < rotary spade < mouldboard plough (**Figure 2**). However, a risk of mouldboard ploughing, or rotary spading is wind erosion which can be minimised by ploughing during a wet season and immediately seeding with a vigorous cereal or cover crop [68]. The duration of the benefits of ameliorating soil water repellence using ST varies between different ST implements. For example, the longevity of the soil inversion with a mouldboard plough has been reported up to more than 10 years [34] and about 3–7 years for the rotary spading [35].

3.1.2 ST for treating subsoil compaction

Approximately 12 million hectares of the south-western agricultural region of WA (75% of arable land) is affected by or susceptible to compaction (**Figure 3**). The annual associated loss of crop yield caused by compaction in WA has been estimated about $AU883 million or $AU36–87 per hectare per year [56]. One of the main objectives of ST is to remove soil compaction allowing plant root growing in the subsoil and accessing soil water and nutrients from the subsoil. Soil penetrometer resistance (also known as soil strength) and bulk density are widely used to measure soil compaction. A soil strength of 2.5 MPa is considered a critical value above which plant root growth is severely arrested [14].

According to Blanco-Canqui and Ruis [70], bulk density and penetrometer resistance did not generally differ between long-term NT and ST systems. However, it is important to note that the lack of soil penetrometer data availability and inconsistency with the measurement was mentioned as a cause of not having a distinct impact of ST

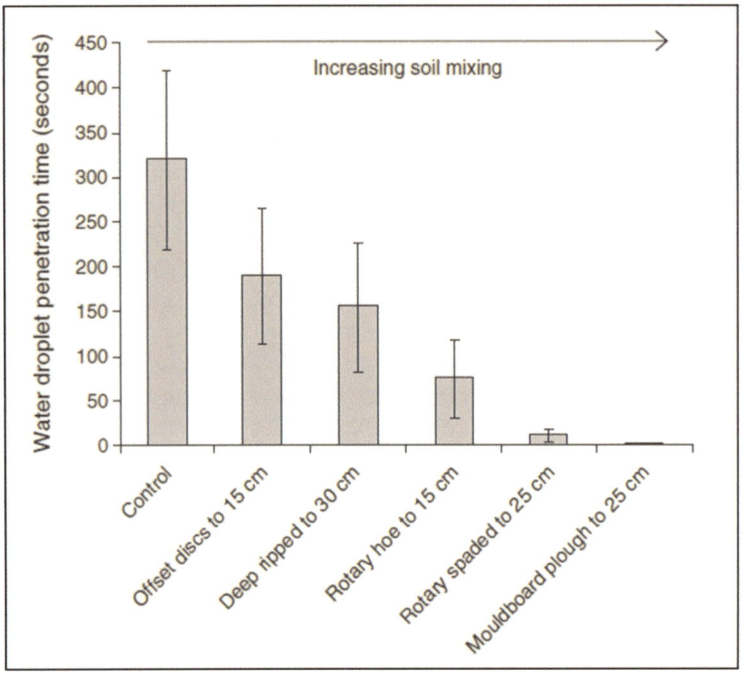

Figure 2.
Impact of one-off cultivation of different intensities on soil water repellence of a Tenosol at Badgingarra as measured in the laboratory using the water-droplet penetration time. Capped lines are standard errors of the mean (n = 4). Adapted from Davies et al. [69].

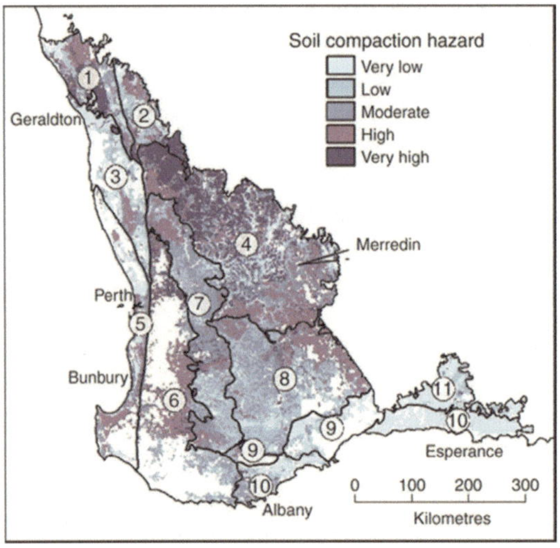

Figure 3.
Status of soil compaction in Western Australia. Source: Parker et al. [56].

on soil compaction [36]. Moreover, the study by Blanco-Canqui and Wortmann [36] used ST implement that addressed only the top 0.20 m of soil which might not alleviate compaction in deeper soil, therefore, deeper tillage was recommended. Azam and Gazey [14] conducted deep tillage (excavation and filling back) in a highly

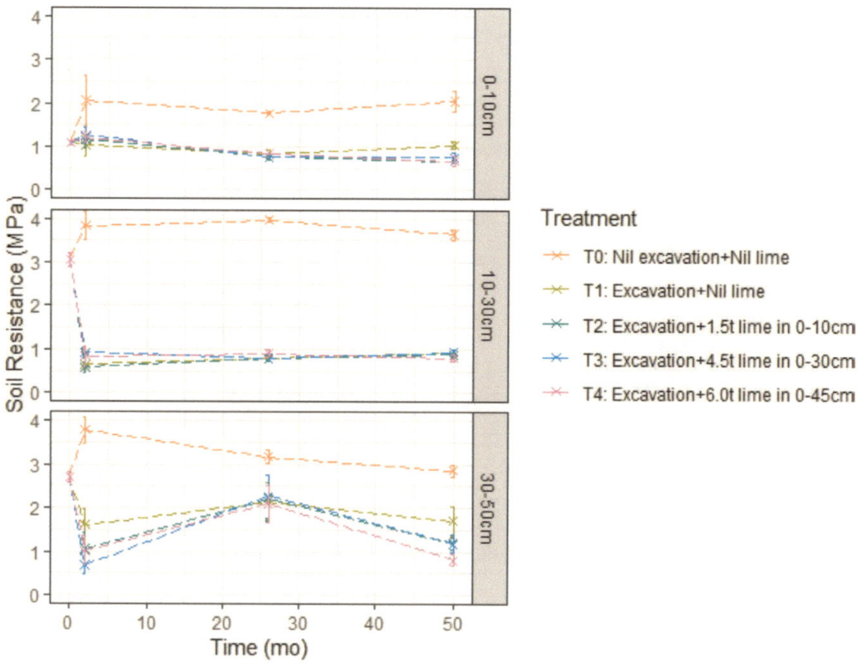

Figure 4.
Effect of strategic tillage (excavation and filling back) on long-term soil resistance. Adapted from Azam and Gazey [14].

compacted soil (a soil strength of 3–4 MPa) in a low rainfall region of Kalannie, WA (35°42'S, 117°29'E) in 2018. They reported deep tillage significantly decreased soil strength (measured at field capacity) and the effect persisted for at least five cropping years (**Figure 4**). Davies et al. [33] reported that deep ripping was the most studied ST practice in southern Australia. Deep ripped soil maintained significantly lower soil strength than the unripped soil for 3–7 years depending on the soil type and rainfall region. Soil with less clay content was naturally recompacted faster than soil with more clay content [33, 56].

3.1.3 ST for treating subsurface soil acidity

In Western Australia (WA), around 50% of the subsoil (0.10–0.30 m) are acidic and have a soil pH (1:5 soil to 0.01 M $CaCl_2$ extract) below the minimum target pH of 4.8 in the subsoil (**Figure 5**). Subsoil acidity is less widely distributed and shallower depths in other southern regions of Australia, but still remains as one of the main soil challenges for the farmers [16]. At low soil pH, the increase in the concentration of toxic forms of aluminium (Al) significantly limits root growth and crop yield [71]. Importantly, soil acidity also often occurs with other soil constraints such as compaction and topsoil water repellence [72].

Several soil amelioration approaches are available for Australian growers to manage subsoil acidity and compaction, the two most widely occurring soil constraints. *Approach 1*: the traditional approach whereby agricultural lime is applied (often multiple applications) on the soil surface. This approach takes many years to increase subsoil pH significantly and does not treat subsoil compaction [14, 16]. *Approach 2*: a

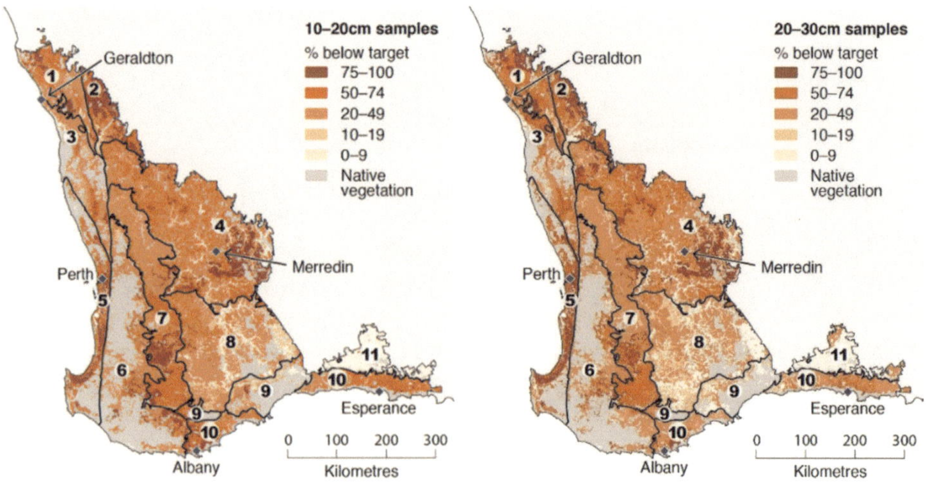

Figure 5.
Status of subsurface soil acidity in Western Australia. Source: Gazey et al. [17].

more recent and increasingly adopted approach that involves the use of ST operations, whereby lime is spread and incorporated at deeper soil depths while loosening compacted soil by a deep ripper or overcoming water-repellent soils by inversion ploughing or rotary spading [33]. Literature suggests that physical tillage operations to treat compaction and non-wetting soils can opportunistically be used to incorporate lime [67]. However, such soil amelioration practices are found to remediate soil acidity only partially, hence yield responses can be variable as observed from various long-term field trials [34]. Scanlan et al. [73] suggested that if a tillage operation incorporates lime to the depths where the soil pH constraint occurs, an immediate payback on lime and cultivation might be possible.

From a long-term (25 years) field experiment with multiple rates (0–8.5 t/ha) and applications of lime, with or without ST, in Wongan Hills, WA (30.°85'S, 116°74′E), soil samples were collected in 2019 and pH was measured [74]. The result showed that the strategic deep tillage was effective in increasing the subsoil pH to the minimum target pH level of 4.8 (in 0.10–0.20 m and 0.20–0.30 m soil depths) within 3 months of applying ST compared to the NT practices, especially where higher rates of lime was applied (**Figure 6**). However, the ST did not show any significant negative effect in changing soil pH of the topsoil (0–0.10 m) compared to the NT [74].

The amelioration of soil acidity will consequently decrease the toxic form Al concentration. In fact, when soil pH (measured in 0.01 M $CaCl_2$) increases to >4.5 pH units, the toxic form of Al almost disappears. An experiment by Azam and Gazey [75] in Kalannie, WA (35°42'S, 117°29′E) reported that lime incorporation increased soil pH to the recommended pH level of 5.5 in the surface and 4.8 in lime-incorporated subsoil horizons (**Figure 7**). Liming also decreased extractable Al concentration from very toxic range (≥30 mg/kg) to below the critical level for wheat (≤5 mg/kg). Loosening by excavation alone also decreased Al concentration, especially at 0.10–0.20 m depth (**Figure 7**).

The longevity of the effect of the lime application on soil acidity may depend on many factors (such as soil texture, rainfall, and moisture content) which might lure more research on that matter. Based on Australian data for ST, Conyers et al. [76] reported that the effect of the addition and incorporation of 2 to 3 t/ha lime should last for at least 10 years.

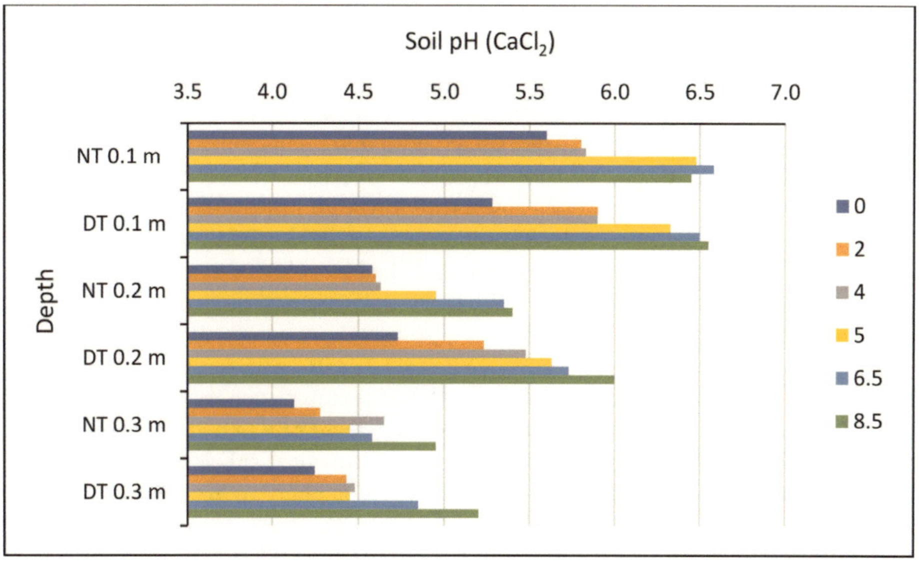

Figure 6.
Soil pH under different liming strategies and tillage treatments in autumn 2019 in loamy sand at Wongan Hills, WA. The NT and DT indicate no-tillage and deep tillage, respectively. Different colours of the Bar indicate lime rates (0–8.5 t/ha). Statistics: The least significant differences (5%) for deep tillage x lime rates at 0–0.10 m depth = 0.36, 0.10–0.20 m depth = 0.60 and 0.20–0.30 m depth = 0.45. Adapted from Azam et al. [74].

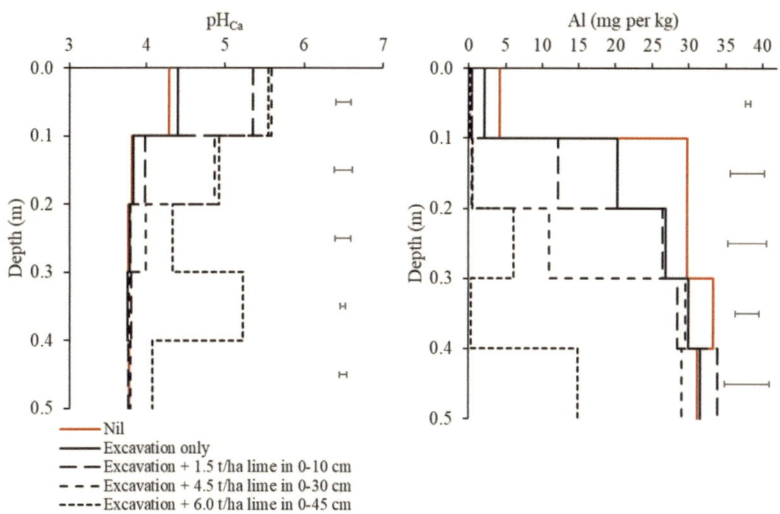

Figure 7.
Effect of lime incorporation on soil pH (left) and extractable Al concentration (right). Horizontal error bars represent the least significant difference at $P \leq 0.05$. Adapted from Azam and Gazey [75].

3.1.4 ST for nutrient redistribution

One of the downsides of conservation agriculture is nutrient stratification along the soil profile. It is a problem for mobile and non-mobile plant nutrients [77]. Broadcasting of immobile nutrients like phosphorus (P) with no or incomplete mixing by tillage will increase nutrient concentration in the surface soil and will pose a greater risk of nutrient loss through runoff and will be greatly responsible for nutrient eutrophication. The leaching of soluble nutrients is another cause of nutrient stratification in the subsurface layers. Soil mixing through tillage may reduce the risk of nutrient loss from topsoil layers [78]. Many studies also reported similar findings of an increase in subsoil nutrient concentration after tillage with a low concentration of nutrients [77–82] due to mixing of top and subsoil [83]. However, the benefit of tillage on nutrient redistribution will disappear over time through improved acquisition by plant roots [84]. McLaughlin et al. [85] rotary hoed a 250 kg/ha superphosphate at the top 0.10 m of soil under an annual pasture and the effect lasted for 7 years.

3.1.5 ST for herbicide-resistant weeds

One of the main factors that accelerated the adoption of conservation agriculture (CA) was the development of herbicides. The long-term practice of CA resulted in the development of resistance against herbicides by many weed species [86, 87]. The advantage of ST is the physical mixing of soil which is likely to decrease weed emergence and bury weed seeds. The emergence and growth of weeds can be stimulated by shallow tillage, but deep tillage can potentially prevent weed emergence [35, 88]. A recent review by Mia et al. [89] reported that soil inversion with mouldboard plough (0.35–0.45 m operating depth) can decrease weed emergence by 50–99% [35, 90–93]. Soil inversion with a modified one-way disc plough, (0.30–0.40 m operating depth) decreased weed population by up to 90% [94]. However, in a recent study by Edwards et al. [95] reported an increase in the activity of some herbicides and an increase in phytotoxicity following ST due to decreased organic matter. Therefore, when ST is adopted, herbicides and their doses should be carefully selected.

3.2 Effect of ST on plant growth and crop yield

3.2.1 Effect of ST on plant root growth

In the Wheatbelt region of WA, many paddocks have multiple soil constraints such as compaction, water repellence and subsurface soil acidity, resulting in limited and confined root growth within 0.20–0.30 m of the surface [96]. The amelioration of soil constraints such as soil acidity and compaction would have many benefits for plant growth for accessing subsoil moisture and nutrients [72]. Azam and Gazey, 2021 showed that amelioration of compaction and acidity in a 25-year NT paddock in Kalannie, WA (35°42’S, 117°29′E) by ST and incorporation of lime significantly enhanced root growth up to 0.70 m soil depth compared to control treatment (where root reached 0.30 m below the soil surface, **Figure 8**).

3.2.2 Effect of ST on crop yield

The amelioration of soil compaction, acidity, and other soil and agronomic constraints using ST and agricultural lime is expected to increase crop yield. The level of yield response from the adoption of ST will depend on whether a single or multiple constraints are addressed in a single intervention. Azam et al. [74] demonstrated using a Mouldboard (MBP) plough only for amelioration water repellent soil generated significant yield response for 7 years. However, incorporation of 2 t/ha lime using a MBP to a depth of 0.25–0.30 m significantly increased yield after 7 years and has continued to have a yield advantage over the next 7 years, compared to the MBP-alone and the

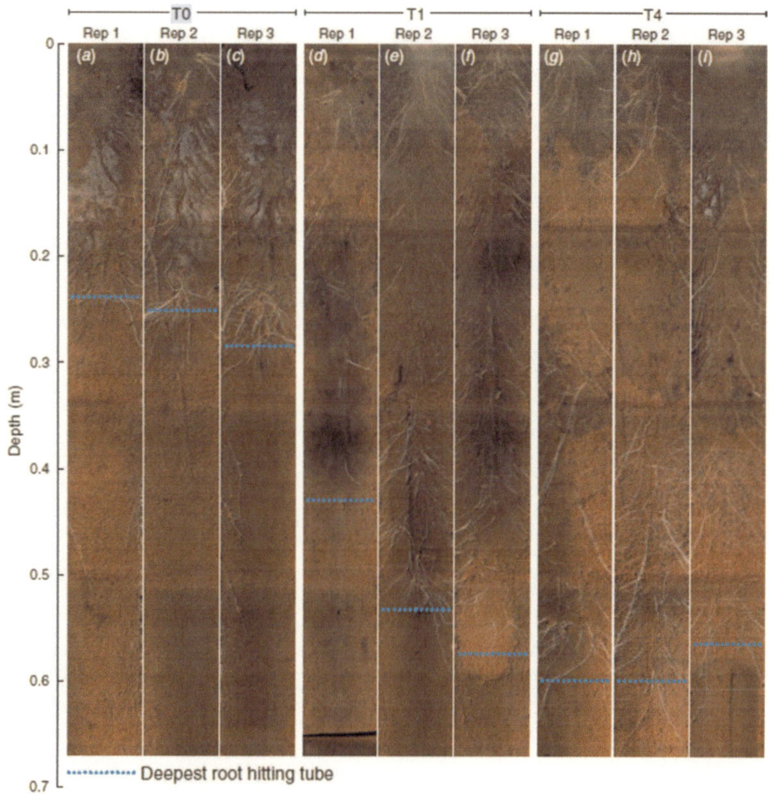

Figure 8.
*Wheat (*Triticum aestivum *L.) rooting depth and architecture in deep loamy sand at Kalannie, WA, 56 days after sowing under control (T0; a–c), removal of compaction only (T1; d–f) and compaction plus deep incorporation of lime (T4; g–i). Wheat root architecture for T0, T1 and T4 was imaged repeatedly in situ using a 360° scanner (CI-600, CID bio-science, camas, WA, USA) inserted in clear glass tubes (Rhizotron tubes, ICT international, Armidale, NSW). Photo adapted from Azam and Gazey [14].*

control [74]. Yield response will also depend on the severity and depth of soil constraints as well as the depth of ST. Davies et al. [33] 2019 reported an increasing yield response with increasing depth of deep ripping (**Figure 9**). However, a recent review of the effect of ST on crop yield by Blanco-Canqui and Wortmann [36] reported that about 80% of the studies (12 studies including 31 soils across different regions of Australia and worldwide) showed no effect, 15% of cases increased the yield. In comparison, there was a yield penalty in 5% of cases. The reason for such variable results was speculated as the variability of site-crop-year-management-specific. Another reason could be the use of shallow tillage activity (only the top 0.20 m of the soil), which does not ameliorate deeper subsoil compaction and acidity and allow plant root to explore further. Such shallow tillage operation might damage to the soil structure and increase the loss of soil water through enhanced evaporation rates [4].

A deep lime incorporation (re-engineering) experiment (see details [74]) was established in Kalannie (30°25’S, 117°17′E) in 2018 using 0, 1.5, 4.5 and 6.0 t/ha lime and monitored for three seasons. Five soil amelioration treatments were included, comprising an untreated control, removal of compaction only and removal of both compaction and acidity (by incorporating lime at 0–0.10, 0–0.30 and 0–0.45 m depths). In this experiment at Kalannie, the combined removal of compaction and

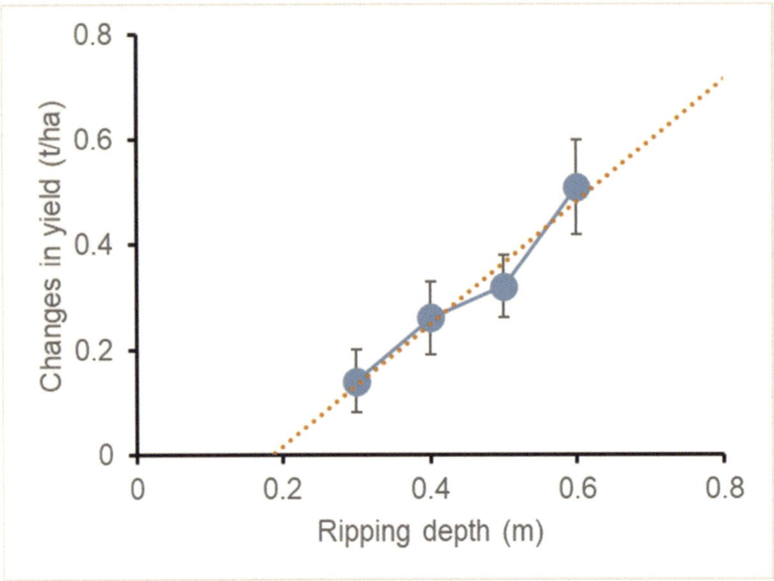

Figure 9.
Relationship between ripping depths and changes in grain yield. The number of data points (comparisons) with 0.30, 0.40, 0.50, and 0.60 m ripping depths are 32, 29, 24, and 14, respectively. Adapted from Davies et al. [33].

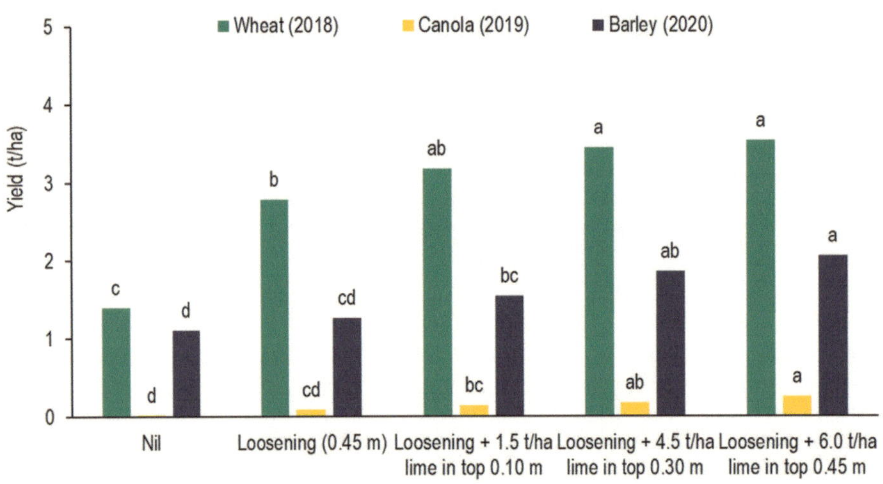

Figure 10.
Grain yield responses due to soil re-engineering in Kalannie. Different letters indicate significant difference in re-engineering experiments at $P \leq 0.05$. Adapted from Azam and Gazey [74].

acidity doubled the grain yield (**Figure 10**). Removal of compaction alone increased the yield of wheat but not of canola or barley. Grain yield exceeded the water-limited potential by 33–56% (as calculated using the French and Schultz equation [97] due to the amelioration of multiple constraints under standard agronomic practice [74]. Deep soil re-engineering allowed plants to produce 0.60–0.65 m deep root systems, while they were only 0.20–0.25 m deep for control [74]. Significant and rapid improvements in root growth and yield were achieved due to the increase in subsoil pH and a uniform and sustained decrease in soil resistance [74, 75].

4. Economic benefits of ST

Currently the occurrence of multiple interacting soil and agronomic constraints is costing Australian farmers millions of dollars due to loss in crop productivity [54–57]. ST can manage these soil and agronomic challenges. However, use of ST always incurs an additional cost. The cost of tillage operation depends on many factors such as type of tillage equipment, type of tillage operation; deep ripping or soil mixing, strategic deep tillage, constraint type; single constraint or multiple constraints [33]. Yield response depends on crop type, soil productivity, rainfall, and other management factors. In calculating the economic benefits of ST, the longevity of operation for specific or multiple interacting constraints needed to be considered.

Among different tillage tools, deep ripping is mostly expensive in southern Australia, ranging from $AU 45 to 1400 and depending on the type of tillage tool and constraints that need to be fixed (**Table 1**). Soil mixing is the cheapest one due to its shallow tillage depth capacity ranging from (0.20–0.40 m). The cost of soil inversion is higher than soil mixing ranging from $AU 40 to 150 but would address more soil constraints.

The existing literature showed that soil mixing or inversion is the cheapest tillage method with maximum yield response and long-lasting benefits compared to the deep ripping method of tillage (**Table 2**). Although deep ripping methods show less economic return in the current literature, it requires more study and time to justify their contribution to yield. Because deep ripping up to 0.70 m is a new concept that requires careful attention to build the necessary equipment to reduce the operation cost at a cheaper rate. Deep ripping using a 'soil re-engineering' approach showed a greater yield increase potential in recent years (**Figures 8** and **9**) and require a longer observation before concluding its effect on longevity. Moreover, deep ripping in ST could address multiple soil constraints following the soil re-engineering approach.

Tillage	Tillage tool	Tillage depth (m)	Soil constraints addressed*	Cost (approx) ($AU/ha)
Deep ripping	Ripper	0.3–0.7	1, 2	$45–100
	With topsoil slotting		1, 2, 3, 4	$55–120
	Subsoil placement		1, 2, 3, 5	$300–1400
Soil mixing	Offset discs	0.2–0.3	1, 3, 6	$50–70
	One pass tillage	0.3–0.35	1, 3	$70–100
Soil inversion	Rotary spader	0.35–0.4	1, 3, 5, 6	$120–150
	Mouldboard plough	0.35–0.45	1, 5, 6, 7	$100–150
	Modified one-way disc plough	0.3–0.4	1, 5, 6, 7	$40–60

**Soil constraints included (1) compaction, (2) Hardpans, (3) subsoil acidity (4) subsoil sodicity, (5) subsoil fertility, (6) water repellence, and (7) weeds.*
Adapted from Davies et al. [33].

Table 1.
Cost involvement in selecting tillage techniques and methods to specific/group of soil constraints.

Tillage method	Features	Yield responses	References
Deep ripping	• More effective in sandy soil than clay soil • Benefit last for 3 seasons	*Wheat* NSW: 33% SA: 10–23% VIC: 23–25% WA: 20–47%	[98]
Deep ripping with amendments	• Gypsum, nutrients, wheat straw + nutrient, chicken manure	VIC: 12–16%	[33]
Soil mixing or inversion	• Benefits last for 10 years	*WA: In Cereal* 1st and 2nd year: 56–86% 3rd & after 11–49%	[33]
		SA Field research Yield increase: 200%	[99]

NSW = New South Wales, SA = South Australia, VIC = Victoria, and WA = Western Australia.

Table 2.
Yield response from different tillage methods practised in various regions of southern Australia.

5. Future research direction

Continuous practices of conservation agricultural (CA) has created a new generation of soil and agronomic challenges that can be strategically managed using one-off deep tillage (ST) and then returning to CA in a no-till cropping. ST is likely to offer a sustainable solution to the soil and agronomic constraints that arise from long-term CA. However, the longevity of the ST in an NT system after fixing single or multiple soil constraints needed to be explored further.

Soil re-engineering is a new approach that signifies the importance of ST, however, identifying the right soil re-engineering approaches for treating multiple interacting soil constraints needs further research before being ready for adoption at farmer scale.

Developing the next generation soil amelioration machinery for optimum amelioration outcome is required. The deep tillage technique requires heavy machinery which would involve an elevated cost to the farmers for agricultural activities. Therefore, new machinery options must be explored to attain a cheaper investment cost for agricultural activities.

Soil water holding capacity (also known as crop upper limit) is usually determined by soil textural classes and by soil organic matter. Under conservation agriculture, soil texture cannot be readily changed, therefore, soil water holding capacity remains unchanged. Using strategic tillage implements such as deep-ripper or delver in soil amelioration can bring deep subsoil clay (especially in duplex soil) to the upper sandy layers (where most crop roots grow) and improve the soil water holding capacity. In re-engineered soil, a prescribed amount of clay can be added to the sandy layer, significantly increasing water holding capacity while maintaining adequate water infiltration. Plant available water (the amount of water a soil can hold after free drainage minus the amount of water at the crop's lower limit) in ameliorated soil will differ from the constrained soil due to restricted root growth. Therefore, an improved understanding of the relationship between plant available water, plant growth and crop yield responses for strategically tilled soil is necessary.

Acknowledgements

We thank the Government of Western Australia and the Grains Research and Development Corporation for investing in various soil research projects under the Soil Constraints West banner. We are grateful to Chris Gazey and Dr. Stephen Davies for pioneering the management of multiple soil constraints in Western Australian agriculture and the many challenging and productive conversations in this area. We thank Daron Malinowski, Richard Bowles, Hasinur Rahman, Alistair Hall for helping in these research projects and other technical staff from DPIRD's research support operations.

Conflict of interest

The authors declare no conflicts of interest.

Author details

Gaus Azam*, Md. Shahinur Rahman and Kanch Wickramarachchi
Department of Primary Industries and Regional Development, Northam, Western Australia, Australia

*Address all correspondence to: gaus.azam@dpird.wa.gov.au

References

[1] Emerson WW. Inter-particle bonding. In: Soils: An Australian Viewpoint. Div. Of Soils, CSIRO, Melbourne. London, England: Academic Press; 1983. pp. 477-498

[2] Melland AR, Antille DL, Dang YP. Effects of strategic tillage on short-term erosion, nutrient loss in runoff and greenhouse gas emissions. Soil Research. 2016;**55**:201-214. DOI: 10.1071/SR16136

[3] Dang YP, Balzer A, Crawford M, Rincon-Florez V, Liu H, Melland AR, et al. Strategic tillage in conservation agricultural systems of North-Eastern Australia: Why, where, when and how? Environmental Science and Pollution Research. 2018;**25**:1000-1015. DOI: 10.1007/s11356-017-8937-1

[4] Betti G, Azam G. Topsoil Evaporation in Water Repellent Soil Affected by Tillage and Claying: Preliminary Case-study Results. In: Proceedings of Soil Science Australia WA Branch Conference. Perth, Western Australia: University of Western Australia; 2019. Available from: https://library.dpird.wa.gov.au/cgi/viewcontent.cgi?article=1077&context=ssswa_conf

[5] Swaminathan C, Sobhana E, Kannan P, Yassin MM. Principles, positives and limitations of conservation agriculture: A review. Agricultural Reviews. 2022;**43**:410-418. DOI: 10.18805/ag.R-2166

[6] Food and Agriculture Organization. Conservation Agriculture. 2022. Available from: https://www.fao.org/3/cb8350en/cb8350en.pdf [Accessed: June 15, 2023]

[7] Kirkegaard J, Kirkby C, Oates A, van der Rijt V, Poile G, Conyers M. Strategic tillage of a long-term, no-till soil has little impact on soil characteristics or crop growth over five years. Crop and Pasture Science. 2020;**71**:945-958. DOI: 10.1071/CP20334

[8] Fischer RA, Hobbs P. Tillage: global update and prospects. In: Pratley J, Kirkegaard J, editors. Australian agriculture in 2020: from conservation to automation: Graham Centre for Agricultural Innovation. Wagga Wagga, NSW, Australia: Charles Sturt University; 2019. p. 3-19

[9] Kassam A, Friedrich T, Derpsch R. Global spread of conservation agriculture. International Journal of Environmental Studies. 2019;**76**:29-51. DOI: 10.1080/00207233.2018.1494927

[10] Llewellyn R, Ouzman J. Conservation agriculture in Australia: 30 years on. In: Pratley J, Kirkegaard J, editors. Australian agriculture in 2020: from conservation to automation: Graham Centre for Agricultural Innovation. Wagga Wagga, NSW, Australia: Charles Sturt University; 2019. p. 21-31

[11] Umbers A. Farm Practices Survey Report 2016. 2017. Available from: https://grdc.com.au/resources-and-publications/all-publications/publications/2018/farm-practices-survey-report-2016 [Accessed: June 15, 2023]

[12] Chan KY. Development of seasonal water repellency under direct drilling. Soil Science Society of America Journal. 1992;**56**:326-329. DOI: 10.2136/sssaj1992.03615995005600010054x

[13] Hall DJ, Jones HR, Crabtree WL, Daniels TL. Claying and deep ripping can increase crop yields and profits on water repellent sands with marginal fertility in southern Western Australia.

Soil Research. 2010;**48**:178-187. DOI: 10.1071/SR09078

[14] Azam G, Gazey C. Slow movement of alkali from surface-applied lime warrants the introduction of strategic tillage for rapid amelioration of subsurface acidity in South-Western Australia. Soil Research. 2020;**59**:97-106. DOI: 10.1071/SR19329

[15] Conyers MK, Heenan DP, McGhie WJ, Poile GJ. Amelioration of acidity with time by limestone under contrasting tillage. Soil and Tillage Research. 2003a;**72**:85-94. DOI: 10.1016/S0167-1987(03) 00064-3

[16] Li GD, Conyers MK, Helyar KR, Lisle CJ, Poile GJ, Cullis BR. Long-term surface application of lime ameliorates subsurface soil acidity in the mixed farming zone of South-Eastern Australia. Geoderma. 2019;**338**: 236-246. DOI: 10.1016/j.geoderma.2018.12.003

[17] Gazey C, Andrew J, Griffin E. Soil acidity. In: Report Card on Sustainable Natural Resource Use in Agriculture: Department of Primary Industries and Regional Development, Perth, Western Australia; 2013. pp. 26-43. Available from: https://www.agric.wa.gov.au/sites/gateway/files/Report%20card%20on%20sustainable%20natural%20resource%20use%20in%20agriculture%20-%20full%20report%20-%2025MB.pdf

[18] Paul K, Black A, Conyers M. Development of acidic subsurface layers of soil under various management systems. Advances in Agronomy. 2003;**78**:187-214. DOI: 10.1016/S0065-2113(02)78005-X

[19] Glen DM, Symondson WO. Influence of soil tillage on slugs and their natural enemies. In: Eli Titi A, editor. Soil Tillage in Agroecosystems. Boca Raton: CRC Press; 2003. pp. 207-227. DOI: 10.1201./9781420040609.ch8

[20] Rahman L, Chan KY, Heenan DP. Impact of tillage, stubble management and crop rotation on nematode populations in a long-term field experiment. Soil and Tillage Research. 2007;**95**:110-119. DOI: 10.1016/j.still.2006.11.008

[21] Owen MJ, Walsh MJ, Llewellyn RS, Powles SB. Widespread occurrence of multiple herbicide resistance in Western Australian annual ryegrass (Lolium rigidum) populations. Australian Journal of Agricultural Research. 2007;**58**: 711-718. DOI: 10.1071/AR06283

[22] Powles SB, Yu Q. Evolution in action: Plants resistant to herbicides. Annual Review of Plant Biology. 2010;**61**: 317-347. DOI: 10.1146/annurevarplant-042809-112119

[23] Pierce FJ, Fortin MC, Staton MJ. Periodic plowing effects on soil properties in a no-till farming system. Soil Science Society of America Journal. 1994;**58**:1782-1787. DOI: 10.2136/sssaj1994.03615995005800060029x

[24] Kirkegaard JA. A review of trends in wheat yield responses to conservation cropping in Australia. Australian Journal of Experimental Agriculture. 1995;**35**: 835-848. DOI: 10.1071/EA9950835

[25] Dick WA. Tillage system impacts on environmental quality and soil biological parameters. Soil and Tillage Research. 1997;**41**:165-167. DOI: 10.1016/S0167-1987(96)01093-8

[26] Bolliger A, Magi J, Amado TJC, Neto FS, Ribeiro MDD, Calegari A, et al. Taking stock of the Brazilian “zero-till revolution”: A review of landmark research and farmers’ practice. Advances in Agronomy. 2006;**91**:47-110. DOI: 10.1016/S0065-2113(06)91002-5

[27] Baan CD, Grevers MCJ, Schoenau JJ. Effects of a single cycle of tillage on long-term no-till prairie soils. Canadian Journal of Soil Science. 2009;**89**:521-530. DOI: 10.4141/cjss08041

[28] Giller KE, Witter E, Corbeels M, Tittonell P. Conservation agriculture and smallholder farming in Africa: The heretics' view. Field Crops Research. 2009;**114**:23-34. DOI: 10.1016/j.fcr.2009.06.017

[29] Domínguez A, Bedano JC, Becker AR. Negative effects of no-till on soil macrofauna and litter decomposition in Argentina as compared with natural grasslands. Soil and Tillage Research. 2010;**110**:51-59. DOI: 10.1016/j.still.2010.06.008

[30] Wortmann CS, Drijber RA, Franti TG. One-time tillage of no-till crop land five years post-tillage. Agronomy Journal. 2010;**102**:1302-1307. DOI: 10.2134/agronj2010.0051

[31] Crawford MH, Rincon-Florez V, Balzer A, Dang YP, Carvalhais LC, Liu H, et al. Changes in the soil quality attributes of continuous no-till farming systems following a strategic tillage. Soil Research. 2015;**53**(3):263-273. DOI: 10.1071/SR14216

[32] Nunes MR, da Silva AP, Denardin JE, Giarola NF, Vaz CM, Van Es HM, et al. Soil chemical management drives structural degradation of Oxisols under a no-till cropping system. Soil Research. 2017;**55**:819-831. DOI: 10.1071/SR17063

[33] Davies S, Armstrong R, Macdonald LM, Condon J, Peterson E. Soil constraints: A role for strategic deep tillage. In: Pratley J, Kirkegaard J, editors. Australian Agriculture in 2020: From Conservation to Automation. Wagga Wagga, NSW, Australia: Agronomy Australia and Charles Stuart University; 2019. p. 117-135. Available from: https://researchoutput.csu.edu.au/ws/portalfiles/portal/35078495/35078149_peer_review.pdf

[34] Davies S, Boyes T, Reynolds C, Petersen L, Walker J, Fulwood T, et al. Understanding the impact of a comprehensive range of deep tillage practices on crop performance and soil properties for two water repellent soil types. In: Proceedings of GRDC Grains Research Updates, 26-27 February 2018. Perth, WA, Australia: Grains Research and Development Corporation; 2018. Available from: https://grdc.com.au/resources-andpublications/grdc-update-papers/tabcontent/grdc-update papers/2018/02/understanding-the-impactof-deep-tillage-practices

[35] Roper MM, Davies SL, Blackwell PS, Hall DJ, Bakker DM, Jongepier R, et al. Management options for water-repellent soils in Australian dryland agriculture. Soil Research. 2015;**53**:786-806. DOI: 10.1071/SR14330

[36] Blanco-Canqui H, Wortmann CS. Does occasional tillage undo the ecosystem services gained with no-till? A review. Soil and Tillage Research. 2020;**198**:104534. DOI: 10.1016/j.still.2019.104534

[37] Blanco-Canqui H, Wienhold BJ, Jin VL, Schmer MR, Kibet LC. Long-term tillage impact on soil hydraulic properties. Soil and Tillage Research. 2017;**170**:38-42. DOI: 10.1016/j.still.2017.03.001

[38] Uusitalo R, Lemola R, Turtola E. Surface and subsurface phosphorus discharge from a clay soil in a nine-year study comparing no-till and plowing. Journal of Environmental Quality. 2018;**47**(6):1478-1486. DOI: 10.2134/jeq2018.06.0242

[39] Rakkar MK, Blanco-Canqui H. Grazing of crop residues: Impacts on soils and crop production. Agriculture, Ecosystems & Environment. 2018;**258**: 71-90. DOI: 10.1016/j.agee.2017.11.018

[40] Blanco-Canqui H, Wortmann CW. Crop residue removal and soil erosion by wind. Journal of Soil and Water Conservation. 2017;**72**:97A-104A. DOI: 10.2489/jswc.72.5.97A

[41] Mendez MJ, Aimar SB, Buschiazzo DE. PM10 emissions from aggregate fractions of an Entic Haplustoll under two contrasting tillage systems. Aeolian Research. 2015;**19**: 195-201. DOI: 10.1016/j.aeolia.2015.01.009

[42] Li H, Tatarko J, Kucharski M, Dong Z. PM2. 5 and PM10 emissions from agricultural soils by wind erosion. Aeolian Research. 2015;**19**:171-182. DOI: 10.1016/j.aeolia.2015.02.003

[43] Nunes MR, Denardin JE, Pauletto EA, Faganello A, Pinto LF. Effect of soil chiseling on soil structure and root growth for a clayey soil under no-tillage. Geoderma. 2015;**259**: 149-155. DOI: 10.1016/j.geoderma.2015.06.003

[44] DeLaune PB, Sij JW. Impact of tillage on runoff in long term no-till wheat systems. Soil and Tillage Research. 2012;**24**:32-35. DOI: 10.1016/j.still.2012.04.009

[45] Quincke JA, Wortmann CS, Mamo M, Franti T, Drijber RA, Garcia JP. One-time tillage of no-till systems: Soil physical properties, phosphorus runoff, and crop yield. Agronomy Journal. 2007a;**99**:1104-1110. DOI: 10.2134/agronj2006.0321

[46] Celik I, Gunal H, Acar M, Acir N, Bereket Barut Z, Budak M. Strategic tillage may sustain the benefits of long-term no-till in a vertisol under Mediterranean climate. Soil Tillage and Research. 2019;**185**:17-28. DOI: 10.1016/j.still.2018.08.015

[47] Kettler TA, Lyon DJ, Doran JW, Powers WL, Stroup WW. Soil quality assessment after weed-control tillage in a no-till wheat–fallow cropping system. Soil Science Society of America Journal. 2000;**64**(1):339-346. DOI: 10.2136/sssaj2000.641339x

[48] Díaz-Zorita M, Grove JH, Murdock L, Herbeck J, Perfect E. Soil structural disturbance effects on crop yields and soil properties in a no-till production system. Agronomy Journal. 2004;**96**(6):1651-1659. DOI: 10.2134/agronj2004.1651

[49] Spargo JT, Alley MM, Follett RF, Wallace JV. Soil carbon sequestration with continuous no-till management of grain cropping systems in the Virginia coastal plain. Soil and Tillage Research. 2008;**100**:133-140. DOI: 10.1016/j.still.2008.05.010

[50] Zhao X, Xue JF, Zhang XQ, Kong FL, Chen F, Lal R, et al. Stratification and storage of soil organic carbon and nitrogen as affected by tillage practices in the North China plain. PLoS One. 2015;**10**:e0128873. DOI: 10.1371/journal.pone.0128873

[51] Quincke JA, Wortmann CS, Mamo M, Franti T, Drijber RA. Occasional tillage of no-till systems: Carbon dioxide flux and changes in total and labile soil organic carbon. Agronomy Journal. 2007;**99**:1158-1168. DOI: 10.2134/agronj2006.0317

[52] Wortmann CS, Quincke JA, Drijber RA, Mamo M, Franti T. Soil microbial community change and

recovery after one-time tillage of continuous no-till. Agronomy Journal. 2008;**100**:1681-1686. DOI: 10.2134/agronj2007.0317

[53] Garcia JP, Wortmann CS, Mamo M, Drijber R, Tarkalson D. One-time tillage of no-till: Effects on nutrients, mycorrhizae, and phosphorus uptake. Agronomy Journal. 2007;**99**:1093-1103. DOI: 10.2134/agronj2006.0261

[54] Barrett-Lennard E, Hall D, Parker W, Munir R. Soil quality 8: Sodic and alkaline soil. In: Hoyle F, Murphy D, Gazey C, Maddern R, Gazey P, editors. The Soil Quality eBooks Series. Perth, WA, Australia: SoilsWest; 2022. p. 53

[55] Davies S, Fisher J, Roper M, Murphy D. Soil quality 7: Soil water repellence. In: Hoyle F, Murphy D, Gazey C, Maddern R, Gazey P, editors. The Soil Quality eBooks Series. Perth, WA, Australia: Soils West; 2022. p. 48

[56] Parker W, Isbister HF, Leopold M. Soil quality 6: Soil compaction. In: Hoyle F, Murphy D, Gazey C, Maddern R, Gazey P, editors. The Soil Quality eBooks Series. Perth, WA, Australia: SoilsWest; 2021. p. 63

[57] Gazey C, Azam G, Clausen J, Rengel Z. Soil quality 4: Soil acidity. In: Murphy D, Hoyle F, Boggs G, Gazey P, editors. The Soil Quality eBooks Series. Perth, WA, Australia: SoilsWest; 2019. p. 48

[58] Van Gool D. Identifying soil constraints that limit wheat yield in South-West Western Australia. In: Resource management technical report 399: Department of Primary Industries and Regional Development, Perth, Western Australia;2016. Available from: https://www.agric.wa.gov.au/managing soils/soil-constraints-limit-wheat-yield-southwest-western-australia [Accessed 2023-06-15]

[59] Bisdom EB, Dekker LN, Schoute JF. Water repellency of sieve fractions from sandy soils and relationships with organic material and soil structure. Geoderma. 1993;**56**:105-118. DOI: 10.1016/0016-7061 (93) 90103-R

[60] Harper RJ, McKissock I, Gilkes RJ, Carter DJ, Blackwell PS. A multivariate framework for interpreting the effects of soil properties, soil management and landuse on water repellency. Journal of Hydrology. 2000;**231**:371-383. DOI: 10.1016/S0022-1694(00)00209-2

[61] Blanco-Canqui H, Lal R. Extent of soil water repellency under longterm no-till soils. Geoderma. 2009;**149**:171-180. DOI: 10.1016/j.geoderma.2008.11.036

[62] Roper MM, Ward PR, Keulen AF, Hill JR. Under no-tillage and stubble retention, soil water content and crop growth are poorly related to soil water repellency. Soil and Tillage Research. 2013;**126**:143-510. DOI: 10.1016/j.still.2012.09.006

[63] Matějková Š, Šimon T. Application of FTIR spectroscopy for evaluation of hydrophobic/hydrophilic organic components in arable soil. Plant, Soil and Environment. 2012;**58**:192-195. DOI: 10.17221/317/2011-PSE

[64] Nadav I, Tarchitzky J, Chen Y. Soil cultivation for enhanced wastewater infiltration in soil aquifer treatment (SAT). Journal of Hydrology. 2012;**470**: 75-81. DOI: 10.1016/j.jhydrol.2012.08.013

[65] Steenhuis TS, Hunt AG, Parlange JY, Ewing RP. Assessment of the application of percolation theory to a water repellent soil. Soil Research. 2005;**43**:357-360. DOI: 10.1071/ SR04093

[66] Davenport D, Hughes B, Davies S, Hall D. Spread, delve, clay, invert: A best

practice guide to the addition of clay to sandy soils. WA, Australia: Grains Research and Development Corporation; 2011. Available from: https://cdn.environment.sa.gov.au/landscape/docs/mr/spread-delve-spade-invert-rep.pdf [Accessed: June 15, 2023]

[67] Davies SL, Blackwell PS, Bakker DM, Scanlan CA, Gazey C, Hall DJ, et al. Deep soil cultivation to create improved soil profiles for dryland crop production. In: Proceedings of Society for Engineering in Agriculture (SEAg) Conference: Innovative Agricultural Solutions for a Sustainable Future, 22-25 September 2013. Mandurah, WA, Australia: Engineers Australia, Barton, ACT; 2013. p. 410-420

[68] Davies S, Blackwell P, Bakker D, Scanlan C, Roper M, Ward P. Developing and assessing agronomic strategies for water repellent soils. In: Proceedings of Agribusiness Crop Updates, 28-29 February 2012. Perth, WA, Australia: Grains Research and Development Corporation and Department of Agriculture and Food Western Australia; 2012. p. 71-77

[69] Davies S, Newman P, Best B. Impact of soil inversion, soil dilution and claying on nonwetting sandplain soils. In: Proceedings of Agribusiness Crop Updates, 25 February 2010. Perth, WA, Australia: Western Australian Agriculture Authority; 2012. p. 223-228

[70] Blanco-Canqui H, Ruis S. No-tillage and soil physical environment. Geoderma. 2018;**326**:164-200. DOI: 10.1016/j.geoderma.2018.03.011

[71] Kopittke PM, Moore KL, Lombi E, Gianoncelli A, Ferguson BJ, Blamey FP, et al. Identification of the primary lesion of toxic aluminum in plant roots. Plant Physiology. 2015;**167**:1402-1411. DOI: 10.1104/pp.114.253229

[72] Van Gool D. Identifying soil constraints that limit wheat yield in south-west Western Australia. In: Resource Management Technical Report 399. Perth, Western Australia: Department of Primary Industries and Regional Development; 2016. Available from: https://www.agric.wa.gov.au/managing-soils/soil-constraints-limit-wheat-yield-south-west-western-australia [Accessed: June 15, 2023]

[73] Scanlan C, Brennan R, Sarre G. Shifting investment from nutrients to lime and cultivation on acid soils: Is an immediate payback possible? In: Proceedings of Agribusiness Crop Updates (ABCU' 14). Perth, WA, Australia: ABCU; 2014, 2014. pp. 1-6

[74] Azam G, Gazey C, Davies S, Scanlan C. Mind the depth of soil amelioration to surpass the yield gap in Western Australia. In: Proceedings of the 20th Agronomy Australia Conference; 18-22 September 2022; Toowoomba, Qld, Australia: Australian Society of Agronomy; 2022. Available from: http://agronomyaustraliaproceedings.org/images/sampledata/2022/Soil/ASAazam_g_422s.pdf

[75] Azam G, Gazey C. Re-engineering soil pH protiiles to boost water use efficiency by wheat. In: Proceedings of the Agronomy Australia Conference, 25-29 August 2019. Wagga Wagga, NSW, Australia: Australian Society of Agronomy; 2019. Available from: http://agronomyaustraliaproceedings.org/images/sampledata/2019/2019ASA_Azam_Gaus_351.pdf

[76] Conyers MK, Mullen CL, Scott BJ, Poile GJ, Braysher BD. Long-term benefits of limestone applications to soil properties and to cereal crop yields in

southern and Central New South Wales. Australian Journal of Experimental Agriculture. 2003;**43**(1):71-78. DOI: 10.1071/EA01121

[77] Wolkowski, D. Tillage Effects On Nutrient Stratification and Soil Test Recommendations, New Horizons in Soil Science, Department of Soil Science, University of Wisconsin-Madison. 2006. Available from: https://soilsextension.webhosting.cals.wisc.edu/wp-content/uploads/sites/68/2014/02/2006SoilTestStratification.pdf [Accessed: June 15, 2023]

[78] Vu DT, Tang C, Armstrong RD. Tillage system affects phosphorus form and depth distribution in three contrasting Victorian soils. Soil Research. 2009;**47**:33-45. DOI: 10.1071/SR08108

[79] Bauer PJ, Frederick JR, Busscher WJ. Tillage effect on nutrient stratification in narrow-and wide-row cropping systems. Soil and Tillage Research. 2002;**66**(2): 175-182. DOI: 10.1016/S0167-1987(02)00025-9

[80] Wright AL, Hons FM, Lemon RG, McFarland ML, Nichols RL. Stratification of nutrients in soil for different tillage regimes and cotton rotations. Soil and Tillage Research. 2007;**96**:19-27. DOI: 10.1016/j.still.2007.02.005

[81] Ryan MH, Tibbett M, Lambers H, Bicknell D, Brookes P, Barrett-Lennard EG, et al. Pronounced surface stratification of soil phosphorus, potassium and sulfur under pastures upstream of a eutrophic wetland and estuarine system. Soil Research. 2017;**55**: 657-669. DOI: 10.1071/SR16144

[82] Obour AK, Holman JD, Simon LM, Schlegel AJ. Strategic tillage effects on crop yields, soil properties, and weeds in dryland no-tillage systems. Agronomy. 2021;**11**:662. DOI: 10.3390/agronomy11040662

[83] Scanlan CA, Davies SL. Soil mixing and redistribution by strategic deep tillage in a sandy soil. Soil and Tillage Research. 2019;**185**:139-145. DOI: 10.1016/j.still.2018.09.008

[84] Nash DM, Watkins M, Heaven MW, Hannah M, Robertson F, McDowell R. Effects of cultivation on soil and soil water under different fertiliser regimes. Soil and Tillage Research. 2015;**145**: 37-46. DOI: 10.1016/j.still.2014.08.006

[85] McLaughlin MJ, Baker TG, James TR, Rundle JA. Distribution and forms of phosphorus and aluminum in acidic topsoils under pastures in South-Eastern Australia. Soil Research. 1990; **28**:371-385. DOI: 10.1071/SR9900371

[86] Busari MA, Kukal SS, Kaur A, Bhatt R, Dulazi AA. Conservation tillage impacts on soil, crop and the environment. International Soil and Water Conservation Research. 2015;**3**: 119-129. DOI: 10.1016/j.iswcr.2015.05.002

[87] Smith RG, Ryan MR, Menalled FD. Direct and indirect impacts of weed management practices on soil quality. In: Hatiield J, Sauer T, editors. Soil Management: Building a Stable Base for Agriculture. WI, USA: American Society of Agronomy, Soil Science Society of America; 2011. p. 275-86. DOI: 10.2136/2011.soilmanagement.c18

[88] Chauhan BS, Gill GS, Preston C. Tillage system effects on weed ecology, herbicide activity and persistence: A review. Australian Journal of Experimental Agriculture. 2006;**46**: 1557-1570. DOI: 10.1071/EA05291

[89] Mia MS, Azam G, Nouraei S, Borger C. Strategic tillage in Australian

conservation agricultural systems to address soil constraints: How does it impact weed management? Weed Research. 2023;**63**:12-26. DOI: 10.1111/wre.12564

[90] Colbach N, Roger-Estrade J, Chauvel B, Caneill L. Modelling vertical and lateral seed bank movements during mouldboard ploughing. European Journal of Agronomy. 2000;**13**:111-124. DOI: 10.1016/S1161-0301(00)00069-1

[91] Douglas A, Peltzer SC. Weed management—Balancing people plant Profit. In: Sindel BM, Johnson SB, editors. Proceedings of the 14th Australian Weed Conference, 6-9 September 2004. NSW, Australia: Weed Society of New South Wales; 2004. p. 300-303

[92] Mohler CL, Frisch JC, McCulloch CE. Vertical movement of weed seed surrogates by tillage implements and natural processes. Soil and Tillage Research. 2006;**86**: 110-122. DOI: 10.1016/j.still.2005.02.030

[93] Roger-Estrade J, Colbach N, Leterme P, Richard G, Caneill J. Modelling vertical and lateral weed seed movements during mouldboard ploughing with a skim-coulter. Soil and Tillage Research. 2001;**63**:35-49. DOI: 10.1016/S0167-1987(01)00229-X

[94] Agriculture and Food Western Australia. Weed Seed Wizard. Version 7.1.7. 2020. Available from: https://www.agric.wa.gov.au/weed-seed-wizard-0 [Accessed: April 29, 2020]

[95] Edwards T, Davies S, McDonald M, Hall D, Moore J. Understanding interactions between pre-emergent herbicides and inversion tillage. In: Proceedings of the GRDC Grains Research Updates (GRU'18), 26 February 2018. Perth, Western Australia: Grains Research and Development Corporation; 2018. Available from: https://grdc.com.au/resources-andpublications/grdc-update-papers/tabcontent/grdc-updatepapers/2018/02/interactions-between-preemergent-herbicides-and-inversion-tillage

[96] Reynolds CD, Betti G, Davies SL. Testing options for the management of soil water repellence and subsoil compaction in dry environment. In: Proceedings of National Soil Science Conference (NSSC' 18); 18–23 November 2018. Canberra, ACT, Australia: NSSC; 2018. pp. 490-493

[97] French RJ, Schultz JE. Water use efficiency of wheat in a Mediterranean-type environment. I. the relation between yield, water use and climate. Australian Journal of Agricultural Research. 1984;**35**:743-764. DOI: 10.1071/AR9840765

[98] Armstrong RD, Davies SL, Kirkegaard JA. Deep Ripping Not Appropriate for all Soil Types Deep Ripping Factsheet. WA, Australia: Grains Research and Development Corporation; 2009. Available from: https://grdc.com.au/__data/assets/pdf_file/0028/91738/grdc_fs_deepripping_hr.pdf.pdf [Accessed: July 11, 2023]

[99] Macdonald L, McBeath T, Llewellyn R, Moodie M, Desbiolles J, Saunders C, et al. Underperforming sandy soils—Targeting constraints for cost effective amelioration. In: Proceedings of the GRDC Grains Research Updates (GRU'19); 12–13 February 2019. Adelaide, South Australia: GRU; 2019. pp. 79-84

Chapter 9

Modeling Growth and Yield of Crops Using Different Tillage Systems

Simon Ogbeche Odey and Raphael Segun Bello

Abstract

Tillage, an integral component of crop production systems, affects crops growth and yield. Different experiments conducted by researchers on diverse tillage systems, and proposal of different models for predicting crops output were presented. Estimating plant yield has positive value for sustainable development. The number of factors determining crop productivity makes modeling crop production challenging task. Forecasting crop production is challenging, requiring making inferences on future performance based on past conditions. Farmers focus on yield as cropping commences. Formerly, yield prediction was calculated by analyzing farmer's previous experience on particular crop. Finite model equations are used to predict output of crops during cultivation provided soil physical properties and their growth parameters are revealed. Different tillage systems including conventional, conservative, reduced, zero, mulch, ridge, minimum and strip in predicting crops growth and yield were discussed. Notable models for predicting growth and yield of crops using different tillage systems were highlighted. Yield of crops is estimated during cultivation provided soil physical properties, climate conditions, and growth parameters of crops are known. Modern farming, utilizing model equations for estimating output, using soil physical properties, climate conditions and relevant parameters of tillage systems are recommended for improvement in crop production depending on soil types.

Keywords: soil tillage systems, crop growth, yield, finite modeling, forecasting

1. Introduction

Tillage is part and parcel of crop production systems since inception of agricultural practice. The process of soil preparation was greatly refined with the invention of the first plow and since then, various types of tillage equipment and modalities for preparation and cultivation of seedbed have been designed and implemented. Development and application of modern equipment for tillage practices were introduced in the last one decade throughout the world, giving way to the essence of precision manipulation of soil [1]. Modern tillage systems make efforts to minimize the quantity of tillage operations, the amount of energy used in terms of human, materials and machineries [2], the quantity of dust that is created [3] and the bulk of soil disturbance [4].

 IntechOpen

Tillage practices greatly affect crop growth and yield [5]. It is one of the fundamental agrotechnical operations in agriculture as recorded by [4, 6–10]. It is a known fact that passes of tractor wheels during tillage result in stiffness of the soil that is capable of affecting both the soil properties and subsequent effects on crop development to maturity and fruiting. Odey et al. [11] and Odey [12] revealed that soil compaction has adverse effects on crop production, affecting agricultural produce in the entire globe.

A number of researchers have studied how soil manipulation affects agricultural output generally. Adekiya et al. [13] reiterated that tillage is important for sustainable okra production on arable soil of southwest Nigeria. According to Ozpinar and Isik [14], tillage systems affecting soil properties are conventional, conservative, reduced, zero or direct seeding, mulch, ridge, stale seedbed, minimum and strip [1, 15, 16]. The state of the earth for agricultural practice should have enough water and plant food, including air to enhance development of roots through it [17–25].

Nowadays, concentration of tillage practices has been much on soil conservation as well as on zero tillage method, primarily to limit cone index and erosion of soil, reduce drought and cost of soil manipulation, enhancing biological activities in the soil for increased crop nutrients for abundant crop production, limiting overall carbon-dioxide release, taming the consequences and depletion of the ozone layer [11, 12, 26]. Generally, okra production is influenced by physiochemical conditions of the lithosphere, including the load-bearing capacity of the soil, density, structure, texture, colloids, nitrogen (N), pH, phosphorus (P), carbon (C) and many more.

The aim of this chapter is to present different field experiments conducted by researchers on different tillage systems and also to propose different models for predicting growth and yield of crops, for enhancement of agricultural production.

2. Tillage systems

Tillage is the physical manipulation of soil using tillage equipment to cut, lift, turn, pulverize and break soil clogs in order to prepare the soil for planting. Tillage is broadly classified into primary and secondary. Primary tillage is the initial manipulation done on the soil using primary tillage equipment such as plows to break the soil. While secondary tillage operations are subsequent manipulations done on the soil to further prepare the soil for planting. Depending on the type of seedbed required by the farmer, the type of crops to be planted, soil condition, cost of tillage and the availability of equipment, farmers can decide to adapt different types of tillage systems, as explained in the following paragraph. Different researchers/authors have revealed their thoughts on different tillage practices, notable among them are Rasmussen [19]; MWPS (MidWest Plan Service) [20]; Owen [21]; ASAE (American Society of Agriculture and Biological Engineers) [22]; Leonard et al. [23]; Hedayatipoor and Alamooti [24] and Odey [25].

2.1 Conventional tillage

In conventional tillage, operation plowing and harrowing are carried out more than once before planting of crops. This system of tillage has been historically and traditionally used to prepare a seedbed and produce a given crop [20, 22]. Thus, in conventional tillage practice, less than 30% of the soil surface is incorporated by crop remains after cultivation.

2.2 Conservative tillage

Conservative tillage is a tillage practice that incorporates at least 30% of the crop residues after preparation and adopted mainly to reduce water erosion among other benefits. Thus, about 1120 kg/ha is mixed with the soil after tillage [21]. Currently, there is a redirection toward shifting to the conservation and no-tillage methods purposely to reduce soil compaction, control erosion, conserve soil and water in the soil, mitigate drought, reduce cost of tillage, increase soil organic matter, boost crop productivity, and to reduce net CO_2 emissions which contribute to global warming [11, 12, 25, 26].

2.3 Reduced tillage

This type of operation allows minimum number of equipment passes along the cultivated area than the conventional tillage. This system employed in agricultural practice enables at least 15% but less than 30% coverage by surface residue after cropping.

2.4 Zero tillage

In this system, the cultivated land is not manipulated throughout the cropping season. However, soil disturbance could result only during fertilizer incorporation and planting using seed drills, openers and coulters.

2.5 Mulch-tillage

Mulch-tillage refers to a practice of leaving 30% and above of crop remains on the top soil to enhance moisture on the soil, protect and reduce excess heat in the soil and generally maintain optimum soil condition. Different tillage equipment and tools are utilized during mulch-tillage, thereby enhancing and maintaining adequate soil water throughout the cropping season [22].

2.6 Ridge-tillage

In this type of system, the agricultural land is not manipulated throughout the season. Thus, fertilizer and seeds are applied on beds and ridges previously made in the last planting season using specialized equipment and tools, such as sweeps, disks and furrow openers. Weeding operation is carried out by the use of herbicides.

2.7 Stale seedbed

Stale seedbed production systems make use of full-width tillage after harvest, sometimes using tillage equipment comparable to the ones utilized during mulch-tillage operation. In this system, ridges or beds are not manipulated but allowed to stale till the next cropping season. Contact herbicides are utilized in weeding. Preparation of land can take place in this system of tillage compared to mulch-tillage system. Thus, no further manipulation is done after the land preparation.

2.8 Minimum tillage

Minimum tillage has been used by the CT Workgroup [4]. It refers to a practice that reduces tillage passes, leading to conservation of fuel for a given crop by at least 40%. The minimum tillage system aimed at meeting up with the 40% and above decrease in the soil disturbance. As reiterated by the researchers in the University of California, 50% of fuel and 72% of time can be saved when utilizing the machine called "Incorpramaster" that operates on only one strip, in contrast to disk plowing and harrowing [2].

2.9 Strip system

In the strip system of tillage practice, tilling is done on the row where crops are to be planted. This is to make sure that crop remains are removed before it is tilled before planting to ensure residue removal, subsoiler usage, and most times allowing for crop land warming and drying.

2.10 Precision tillage

In this type of tillage system, tillage is carried out exactly at the point(s) where it is required. Cost of tillage is greatly reduced since the number of tractor passes is limited; thereby fuel consumption is reduced.

3. Growth and yield of crops

The load-bearing capacity of the soil, density, structure, texture, colloids, N, pH, P, C and many more have effects on agricultural production and output [27]. Different factors responsible for agricultural production output and hazards associated with cropping are available nutrients, including water in the soil, environmental condition, pests and plant diseases. Agriculturists must monitor and control these factors to enable bountiful harvest.

3.1 Factors that affect crop growth and yield

Five important factors affect crop growth and yield are (i) soil fertility, (ii) soil water availability, (iii) climate conditions, (iv) pests and (v) diseases. Out of these, the availability of water in the soil is affected by the type and efficiency of tillage practice adopted. Tillage systems affect soil structure, compaction and water availability for nutrients' utilization. Eighteen essential nutrients contribute to crops' well-being. Their availability in the soil provides assurance for effective development of crops as these essential nutrients are needed in varying quantities, and hence are categorized into macro- and micro-mineral components. The elements in the soil provide essential nutrients toward the growth and yield of crops in terms of development of parts such as the roots, stems, leaves and fruits. These essential nutrients enhance production of hormones and proteins vis-à-vis manufacture of chlorophyll during photosynthesis. Limited supply or lack of any of the elements leads to decrease in crop growth and overall yield.

Thus, the presence of water in the soil has immediate and direct effect on plant growth and yield. Climate and weather conditions have a direct impact on water availability, with scanty or no precipitation leading to drought, causing death of crops, while excess precipitation leads to too much water, which has a negative impact on crops.

Other factors that cause devastating effects on plant growth and yield are the presence of diseases and pests on the crop land. Factors such as the presence of pests and diseases affect crop growth and yield. They are available in different sizes and shapes causing havoc in different ways to crops at varying degrees. Some nematodes also cause damages to plant roots, thereby truncating nutrients' flow abilities of such roots, which then affect the overall growth and yield of the crops.

3.2 Effect of soil compaction on crop growth and yield

Compaction of soil results due to densification of units of soils, thereby contributing to the decrease in their voids. Soane and Ouwerkerk [28] established that compaction leads to compression of the soil particles, leading to an increase in component weight, with considerable volume of air reduction. Soil compression results when the structure can no longer withstand external pressure, leading to a disintegration of soil structural units, soil volume reduction, increase in bulk density, decrease in porosity and soil hydraulic conductivity. As reported by Manor and Clark [29]; Petersen et al. [30]; Wells et al. [31]; and Alameda and Villar [32], soil compaction limits crop growth and yield, by restricting root development and movement of air and water within the soil. Soil compaction in the surface layer can encourage runoff and soil erosion, thus increasing soil and water losses.

Mason et al. [33] reported that the capacity of plant roots to pierce the soil is limited by increase in the load-bearing capacity of the soil terminating completely on 2.5 mega Pascal mark. Aase et al. [34] reiterated that as the penetrometer reading gradually moves toward the 2.0 mega Pascal mark and passes the said limit, the development of crop roots would have revealed retarded values. Therefore, 2.0 MPa has been looked upon as a process of determining hard pan layer of soil Wells et al. [31]. Raper et al. [35] further contributed that the life-threatening limit of resistance to penetration, limiting root development, is between 40 and 50 cm deep into the soil. Thus, subsoiler may be used to penetrate the soil for an easy root growth. Taylor [36], Monroe and Kladivko [37], Mason et al. [33], and Mari and Changying [38] elucidated that hydrostatic force in the growing section of the plant root offers the vigor required to drive the root top and meristematic region along the counterattacking soil. Thus, as long as the turgor is not enough to overwhelm fence battle and resistance of the soil, the growth of that exact root cap stops. Crops cultivated in compacted soils have revealed a smaller number of lateral roots than plants grown under controlled condition. Soil compaction has disadvantages on plants by—elevating the force against the growth of crop roots; interfering with the area of the voids and exasperating plant root ailments [39–53]. Al-Adawi and Reeder [54] reported that subsoiling generally improves growth and yield of all treatments, including control research.

3.3 Measurement of growth and yield parameters in plants

Different researchers have revealed that the growth and yield parameters of crops are measured by gathering data on a regular basis [40, 41, 43–45, 47, 49, 50, 55–58]. Some of the essential properties normally read include crop height, shoot diameter, branches, leaves, leaf area, biomass weight, root development and structure, and threshed weight of soybean/ha.

3.3.1 Leaf area and leaf area index (LAI)

Leaf area is the area confined by a leaf of a plant. It is the total area of the leaf in square meters (m^2). Represented by meter square per meter square, LAI is referred to as the substantial quantity of leaf content existing within a confined area occupied by plants, and its unit is denoted as (m^2/m^2). In geometrical terms, LAI represents the sum of plain side of a leaf area contained in a matured leaf per geographical area of a group of crops. It is generally defined as the ratio of leaf area to land area and is strongly related to crop yield [59–63]. Thus, the system of assessing development association among plants growing together is referred to as LAI. The expression generally used in approximating leaf area index has been used by Agba et al. [64]:

$$\text{Leaf area index, LAI} = \text{N x Y x } A_L \text{ x } (A_P)^{-1} \qquad (1)$$

N = Mean value of plant canopy or leaves
Y = Total number of crops/cultivated land
A_L = Mean value of each leaf area
A_P = Total range of cultivated land

3.3.2 Dry matter (biomass) yield per plant

Dry matter weight (g) of crop is determined at intervals during cultivation. Specific crops are uprooted at random. The uprooted crops are then dried in the oven at 100°C for 72 h. The average weight (g) values of samples selected at random at each plot, which were already dried in an oven, were booked as the appraised crops' biomass.

3.3.3 Grain yield per hectare (t/ha)

Grain yield per hectare (t/ha) is measured by getting yield per hectare of different units of the cropping area using a weighing balance. The mean value of the various trials is then taken.

3.3.4 Mean relative growth rate (MRGR)

Researchers utilize mean relative growth rates (MRGRs) to compare the growth of seedlings that differ in original size. There are several reasons why this system is applied, which are given as follows: (i) so as to eradicate any development alterations relating to size, (ii) in order to fix crops that are characteristically more feasible, (iii) agglomerated growth of different portions of crops is combined, classes and handling differences are equated and finally (iv) computation of MRGR of intermediate components (root and stem) is unswervingly equivalent to MRGR. The method is normally based on the fact that plant development takes place as percentage of the original magnitude of crop which is constant. This is referred to as the law of compound interest. Researchers make use of this scenario even as the proportionate upsurge fluctuates with growing size, hence the law of variable interest. Examining MRGR of plants is among the various systems applied in associating development alterations arising during trial handlings [65–69].

The mean relative growth rate has been used to assess plant development caused by various values of fertilizer application on the land, control of weeds, soil manipulation systems, moisture content of soil, soil bulk density, voids and void ratio, load-bearing capacity of soil, soil erodibility, floods and soil chemical properties [70–73].

According to Paine et al. [69], the MRGR is given by the following expression:

$$MRGR = \frac{1}{\mathrm{W}}\frac{dw}{dt} = \frac{lnW2 - lnW1}{t2 - t1} \tag{2}$$

Where,
w = weight of oven-dried crop
dw = difference in weight of oven-dried crop
dt = time interlude
ln = ordinary logarithm
w1 = original dry weight of seedlings at initial time, t_1
w2 = concluding weight of dry seedlings at time t_2 and
t2–t1 = period of development in days

Applying this method, Alameda and Villar [32] revealed that a 41% momentous upsurge in relative development rate as seedlings of 17 woody crop species were cultivated under reasonably densified soil with 0.1–1.0 mega pascal in a greenhouse.

4. Modeling growth and yield of crops

Forecasting yields of plants has comprehensive consequences for economic considerations, biology and welfare of humans. Different factors that affect crop production cause the idea of modeling cultivation of plants highly cumbersome. Forecasting the output of cropping activity during agricultural practice possesses a high barrier, since the requirement making interpretations on forthcoming performance is dependent on previous circumstances. Manjula and Djodiltachoumy [74] opined that generation of data is the practice of examining data from various standpoints and abstracting the same into beneficial material. Crop output estimation is a very useful agronomic challenge. Agriculturists normally focus on output immediately after a contemplation of agricultural activities is initiated.

4.1 Estimating crop growth and yield using prior data

In the past, output forecasting was computed by investigating the knowledge of the previous agricultural operation. Sangeetha [75] formulated a method toward forecasting crop output by relying on the knowledge obtained from previous statistics of the agricultural field. Some of these data are consideration of the state of the soil, annual rainfall data, temperature, development of crops and so on. She submitted that the suggested model was found to be more applicable for estimating agricultural output than the previous model.

4.2 Forecasting crop growth and yield using regression model

Odey [76] stated that a regression model was formulated for forecasting the maize development and yield on soil densified by tractor passes. In this estimation, maize yield, Ym, was known in advance, provided the number of tractor wheel passes during operation on the agricultural land was recorded. The study was conducted at the experimental farm land of the Agronomy Department in the Obubra Campus of the Cross River University of Technology, during the planting season of 2016. The experiment was carried out with the intention of knowing how machinery traffic affects properties of soil and development and output of maize planted during the said planting season. Removal of vegetation from the farm land was carried out prior to soil manipulation. Three replications of four different treatments were carried out, adding up to a total of 12 distinct land areas for the experiment. The 12 distinct plots were tagged—[A0], representing zero machinery passes, second, [B10], representing 10 machinery passes, [C20], standing for 20 passes of tractor wheels and [D30] representing, 30 passes of farm machinery wheels. The distinct experimental plot has an area of nine square meter. The agricultural field used for the experiment has a total area of 567 square meters. The 12 experimental plots were arranged using the system —Randomized Complete Block Design [RCBD]. Farm tractor having a rated power of 50 kilowatts (kW) was applied in compacting the experimental plots accordingly. Physical and chemical properties were conducted on the soil samples taken before and after treatments of the plots. Maize (FAMMAX-15) seeds were planted. Data on growth and yield parameters were measured.

As recorded by Odey [76], data were analyzed to obtain the relationship between tractor passes, properties of soil and maize growth and yield. More machinery traffic resulted in higher penetrometer readings, soil bulk densities and reduction in moisture of the soil and soil voids. Machinery traffic leads to an increased positive correlation with soil bulk density and negative correlation with soil voids, moisture of soil and development cum output data of the said crop. Thus, soil bulk densities rose up sharply with increased number of machinery passes on the land, whereas the porosity of soil, crop development and output reduced with more number of machinery wheels on the field. Field area with high machinery traffic of 30 passes yielded the minimum output among the 12 plots. Thus, an average output of 2.56 tons per hectare was realized from [A0], while [B10], [C20] and [D30] had corresponding outputs of 2.11, 1.56 and 1.67 tons per hectare accordingly.

Odey [76] further revealed a mathematical expression for estimating maize output prior to gathering operation as the number of machinery passes was equated with soil bulk density, soil voids, moisture content of soil, crop development and final output, using output data as the dependent variable. Thus, the experimental results yielded the following expression for estimating the output of maize:

$$Y_{maize} = 29.34 + 0.012\,TOT - 11.5\,Bd - 0.75\,Ms - 0.25\,Ps + 0.03\,Mh - 0.34\,Mw + 0.37\,Ml \quad (3)$$

Where, Y_{maize} = grain yield, TOT = tractor traffic, Bd = soil bulk density, Ms. = soil moisture, Ps = soil porosity, Mh = maize plant height, Mw = maize stem width and Ml = number of leaves of maize crop.

Odey [76] concluded that the expression mentioned above had an R^2 and standard error of 0.645 and 0.43, respectively. Hence, the need for controlled tractor traffic on farm lands to control its reversed effects on agricultural land and crop development. The researcher further emphasized tractor traffic below 10 passes on sandy loam fields for favorable maize cultivation.

4.3 Estimating crop development and output using modeling equations

Nkakini and Davies [77] formulated an expression for okra tolerance output to soil densification. The study showed an existence of mutual association between the investigated and okra values obtained from the mathematical expression or model values of okra output in acceptance of densification of soil under different machinery wheels on the various experimental fields that were studied. This mathematical expression revealed that okra development and output significantly improved as a result of imposed compaction due to tractor wheel passes. The said experiment was conducted using a total study area measuring 1656 square meters. Clearing and stumping of the field were carried out. Soil tests were carried out on the randomly collected samples of soil before and after manipulation of soil using tillage machineries. The experiment was laid on a randomized complete block design of four replications. Machinery traffic at i = 0, 5, 10, 15 and 20 per replicate was put in place making a total of 20 subplots using a model SWARAJ 978 FE Tractor. Duly certified okra seeds with minimum percentage germination and purity of 85 and 99% were procured and sown accordingly. Data on growth and yield of maize were used in the model development.

4.4 Finite modeling of growth and yield of okra using different tillage systems

Odey [25] conducted a finite modeling of growth and yield of okra using different tillage systems. The field experiment carried out was aimed at modeling okra growth and yield cultivated using different tillage systems. The experiment used three treatments replicated three times in randomized complete block design (RCBD). These tillage systems that were studied were: [A] Conventional-plowing and harrowing, [B] Conservative tillage system and [C] No till method. The three systems of tillage under study had a multiple of 3. There were nine plots in all. An area of 64 square meters was allocated to each of the nine fields. A space of 2 meters separates each of the plots for tractor maneuvering. Planting of okra at distance, 1 m x 0.8 m, was done. The weeds were removed 3 weeks after cropping and an interval of 2 weeks thereafter until gathering fruits after maturity. An assessment of physiochemical properties was carried out in the laboratory on the soil samples taken before and after planting operation.

Odey [25] further explained that values of okra output were taken and then subjected to analysis on IBM SPSS Statistics Version 21 software. The results revealed reversed correlation among development data, output of okra and bulk density of soil. Whereas soil voids had positive correlation with okra development and output, conventional and conservative tillage had higher okra yield compared to no till method. Finite modeled expression with R^2 of 0.934 on okra output studying various soil manipulation methods was formulated, showing forecasted results almost equating with actual output. The researcher recommended conventional and conservative tillage systems for enhancement in the production of okra fruits in sandy loam soil. Thus, a detailed explanation on how Odey [25] conducted the research is given below for an understanding of the reader:

4.4.1 Experimental site and location

The site of the experiment was experimental farm land of the Agronomy Department, Cross River University of Technology, Obubra Campus. This experiment was conducted during the planting season of 2017/2018. The location of the site was at longitude 08°20′00″E and latitude 6°05′00″N. The rainfall data at this location were about 500–1070 mm annually, with a soil type, sandy loam and a temperature range of 21–30°C. The vegetation was a rain forest zone.

4.4.2 Land preparation and experimental design

The vegetation was cleared by spraying systemic herbicide using a portable knapsack sprayer, since the land area was predominantly covered with weeds, such as elephant grass, *Chromolaena odurata* and Centrosema. The dried vegetation was removed after 2 weeks. A randomized complete block design (RCBD) explained earlier was adopted on the field.

4.4.3 Treatments adopted

4.4.3.1 Conventional tillage

Conventional tillage was carried out using Massey Fergusson—MF 435 2WD/4WD tractor with rated power, 50 kW and total weight, 2122 kg which was utilized in plowing and harrowing the plots before planting the seeds.

4.4.3.2 Conservative tillage

Thus, 1120 kg of crop residues was incorporated into the soil per hectare (1120 kg/ha). This was to meet up with the standard 30% of the land area for cropping as specified by ASAE (American Society of Agriculture and Biological Engineers) [22]. Therefore, a total of 7.168 kg of crop remnants was utilized/mixed with each of the 8 m x 8 m plot sizes dedicated for conservative tillage system and harrowed before planting the seeds in accordance with Reddy et al. [78].

4.4.3.3 Zero tillage

In the zero tillage system, the vegetation was cleared and raked. Seeds were planted directly without manipulating the soil, as reiterated by Ruberson and Phatak [16].

4.5 Planting material and germination

Viable seeds of okra (Agwu Early) were soaked for 1 day before planting according to Omran et al. [79]. Then the seeds were planted at 0.02–0.05 m below the soil surface, selecting 2–3 seeds per slit, then, they were later thinned to two stands. The planting distance employed was 1 m x 0.8 m (inter- and intrarow). Germination of grains, growth and yield data were ideal and were determined using the expression after the researcher Agba et al. [64];

$$\text{Germination percentage} = \frac{\text{Number of plants germinated}}{\text{Total number of seeds planted}}\, x\, 100 \tag{4}$$

4.6 Replanting of ungerminated seedlings and pruning

Replanting of ungerminated crops from the nursery raised close to the experimental plots was conducted according to Khan et al. [80].

4.7 Weeding and fertilizer application

Manuel weeding using cutlass was carried out 3 weeks after planting. The process was repeated after an interval of every 2 weeks till beginning of harvest. Application of fertilizer, NPK (nitrogen, phosphorus and potassium), 12: 12: 17 was done after 3 weeks of planting using ring method at the rate of 3.2 kg/experimental plot.

4.8 Data collection

4.8.1 Collection of soil samples

Soil samples were collected at random using soil cores before tillage operations and during growth and maturity of crops on the entire land area (1024 m^2). The samples were collected at depths of 0–15, 16–30 and 31–45 cm using undisturbed cores. Note that the samples were taken from each plot during the development and fruiting periods at the specified points below the soil surface. Soil samples collected were treated accordingly before laboratory experimentation.

4.8.2 Measurement of growth parameters and yield of okra

At 2 weeks after planting, data on the growth rate, such as plant height, width, number of leaves and flowers, were collected. This operation was repeated every 2 weeks interval till maturity and production. Yield of okra data was gathered every 5 days interval as okra pods (matured green pods) were harvested by using knife and hands.

4.8.3 Soil bulk density and porosity

The soil samples were oven-dried at 100°C for 24 h before determining bulk density and porosity using the technique explained by Black and Hartge [81].

4.9 Analysis

4.9.1 Laboratory and statistical analysis

The laboratory analysis of soil samples was done at the Soil Science Laboratory, University of Agriculture, Makurdi. Hydrometer method, as explained by Gee and Bauder [82], was used for determining the particle size. The pH meter with glass electrode was utilized for determining the pH of soil in water, 1:2 of soil:water ratio. Experimentation with flame photometer, as described by Udo et al. [83], was utilized to determine cation exchange, whereas potassium and sodium, calcium and magnesium were determined in the extract by ethylenediaminetetraacetic acid (EDTA) titration. The dichromate wet-oxidation method, as explained by Nelson and Sommers [84], was used for getting organic matter content. Phosphorus content was realized by the Bray-1 step, as described by the researcher Kuo [85]. The available nitrogen was obtained by the micro-Kjeldahl digestion and distillation system, as put forward by Bremmer [86]. Cation exchange capacity (CEC) was determined according to the method explained by Sumner and Miller [87]. The data obtained on growth and yield parameters of okra during the study were subjected to test and analysis using IBM SPSS Statistics Version 21 software. Among the determined components were correlation, regression and analysis of variance (ANOVA).

4.10 Results and discussion

4.10.1 Soil properties of the experimental site

Table 1 shows the result of pretreatment of soil physical and chemical properties of the study site. Thus, textural class of soil is sandy loam having higher percentage of sand particles among all points below the soil surface taken, viz., 0–15, 16–30 and 31–45 in centimeters and the presence of acidic content was noticed in the different depths. Soil within 0–15 cm depths showed more values of organic matter, followed by points 16–30 and 31–45 accordingly, confirming organic matter decreases with soil depth. The soil available N, P content, cation exchange capacity (CEC), Mg and organic matter were found to be minimal, indicating the inherent low fertility status of tropical soils, as reiterated by Ojeniyi [88].

4.10.2 Soil properties as influenced by treatments, growth parameters and yield of okra

Odey [25] further revealed that the structural properties of soil were affected during field experiment (see **Tables 2** and **3**). **Table 2** shows how the different soil manipulation methods influenced physiochemical properties of soil during growth stages of okra. The results indicated that the treatments have a very similar textural class, sandy loam soil, with a higher percentage of sand characteristics showing in the various treatments. Soil pH remained acidic in all the treatments. Organic matter was maximum at 0–15 cm depth in both conventional and conservative tillage than 16–30 cm and 31–45 cm depths apart from zero tillage.

On the other hand, different tillage systems did not lead to rise in soil N, P content, cation exchange capacity (CEC), Mg, Ca and Na in the different depths under study. These results correspond with the findings of Brady and Weil [89], which revealed that soil contains minerals, organic matter, air and water; confirming the submission of Carter [90] who reported that soil textural class is made of clay, sand and silt.

Thus, conventional and conservative soil manipulation methods resulted in increased output of the crop than the output from no till method system. These findings concurred with the submission of Lal [91], which reiterated that various soil manipulation methods have effects on okra output.

4.10.3 Analytical results for properties of soil, development and okra output

According to Odey [25], the experimental results in **Table 4** revealed the analytical results for properties of soil, development and okra output. There is a strong positive

Depths	% sand	% silt	% clay	Textural class	H_2O 1:1	KCl 1:1	Org C%	Org M%	N%	Bray 1 P PPM	Ca mmol/L	Mg mmol/L	K mmol/L	Na mmol/L	CEC
0–15 cm	76.0	12.0	12.0	Sandy loam	5.93	5.11	0.86	1.49	0.092	3.68	3.71	1.58	0.33	0.64	6.50
16–30 cm	75.9	12.9	11.2	Sandy loam	5.91	5.02	0.77	1.33	0.094	3.81	3.69	1.62	0.34	0.70	6.46
31–45 cm	70.2	13.7	16.3	Sandy loam	5.50	4.80	0.66	1.14	0.088	3.9	3.77	1.66	0.30	0.69	6.43
Average	± 3.3	± 0.9	± 2.7		± 0.24	± 0.16	±0.10	±0.18	±0.0	±0.11	±0.04	±0.04	±0.02	±0.03	±0.04

Source: Field data 2017/2018 cropping seasons [25].

Table 1.
Properties of soil at various depths before treatments of agricultural field.

Treatments	Depths	% sand	% silt	% clay	Textural class	H_2O 1:1	KCl 1:1	Org C %	Org M %	N%	Bray 1 P PPM	Ca mmol/L	Mg mmol/ L	Na mmol/L	Na mmol/L	CEC
A	0–15 cm	71.1	13.1	15.8	Sandy loam	6.17	5.35	0.77	1.33	0.079	3.62	3.58	1.63	0.29	0.68	6.11
	16–30 cm	75.4	11.2	13.4	Sandy loam	6.25	5.15	0.74	1.30	0.078	3.46	3.49	1.61	0.29	0.65	6.18
	31–45 cm	76.2	11.5	12.3	Sandy loam	6.00	5.20	0.71	1.24	0.077	3.11	3.44	1.58	0.29	0.66	6.00
Average		±3.2	±0.85	±2.7		±0.1	±0.1	±0.03	±0.05	±0.0	±0.26	±0.07	±0.03	±0.0	±0.02	±0.09
B	0–15 cm	72.2	13.4	15.4	Sandy loam	5.95	5.00	0.80	1.38	0.077	2.86	3.63	1.57	0.30	0.67	6.20
	16–30 cm	76.3	12.4	11.3	Sandy loam	5.86	5.00	0.77	1.33	0.077	3.01	3.58	1.62	0.27	0.64	6.08
	31–45 cm	69.2	14.2	16.6	Sandy loam	5.00	4.50	0.56	0.97	0.075	3.05	3.80	1.70	0.31	0.70	6.75
Average		±3.6	±0.9	±2.8		±0.5	±0.3	±0.1	±0.2	±0.0	±0.1	±0.1	±0.1	±0.0	±0.03	±0.4
C	0–15 cm	75.4	11.6	13.0	Sandy loam	5.92	5.10	0.72	1.24	0.081	3.77	3.41	1.57	0.26	0.64	5.96
	16–30 cm	77.0	12.1	10.9	Sandy loam	6.11	5.90	0.80	1.38	0.079	3.83	3.51	1.60	0.26	0.66	6.21
	31–45 cm	75.0	11.3	13.7	Sandy loam	6.25	5.33	0.88	1.52	0.075	3.82	3.50	1.62	0.27	0.67	6.22
Average		±1.06	±0.4	±1.5		±0.2	±0.4	±0.1	±0.1	±0.0	±0.03	±0.1	±0.03	±0.01	±0.02	±0.2

Source: Odey [25].

Table 2.
Properties of soil as affected by field experiment.

Treatments/ replicates	Soils Bulk density (DB)	Soil porosity	Height of okra (cm)	Width of okra (cm)	No. of leaves/ stand	No. of flowers/ stand	No. of fruits/ stand	Total yield (kg/plot-64 m^2)	Total yield (kg/ha)
A1	0.49	0.82	77	19.78	43.2	17.6	13.6	45.6	7125
A2	0.52	0.80	75.8	17.76	44	6.4	8	40.8	6375
A3	0.54	0.80	80.6	23.84	48	4	9.6	48.8	7625
B1	0.46	0.83	88.6	19.12	56.8	16	16.8	62.4	9750
B2	0.54	0.79	56.6	13.02	45.6	14.4	7.2	29.6	4625
B3	0.46	0.83	81.4	25.08	68.8	19.2	19.2	47.2	7375
C1	0.54	0.80	58	14.24	33.6	9.6	15.2	31.2	4875
C2	0.59	0.78	75	16.54	44	8.8	8.0	36.8	5750
C3	0.52	0.80	68.4	14.08	32.8	8.0	5.6	40.8	6375
Average	±0.04	±0.02	±10.68	±4.26	±11.09	±5.37	±4.83	±10.00	±1562.2

Source: Odey [25].

Table 3.
Average data for density of soil, voids, development and output of okra.

	Bulk density	Porosity	Height	Width	Leaves	Flowers	Total yield
Bulk density	1.000						
Porosity	−1.000	1.000					
Height	−0.5249	0.5249	1.000				
Width	−0.4904	0.4904	0.7707	1.000			
Leaves	−0.6115	0.6115	0.6399	0.7520	1.000		
Flower	−0.6825	0.6825	0.1731	0.2086	0.5799	1.000	
Total yield	−0.6909	0.6909	0.9162	0.6459	0.5731	0.2596	1.000

Table 4.
Analytical values of properties of soil, development and okra output.

relationship between height, width and leaves of okra, showing the correlation between height and width of okra tending to 1 (1.0), and in agreement to that of leaves and width. The findings are related to the assertion of Ariyo and Akenova [92], which expressed a strong affiliation among the development criteria of crops. Hence, there is a strong positive correlation among okra development and output and soil voids. Thus, as densities of voids were 1.000, tallness, breadth, leaf count, flowering and okra output maintain their stand at 0.5248, 0.4904, 0.6115, 0.6825 and 0.6909 accordingly. Whereas, sturdy adverse correlation was displayed between densities of soil, development and okra output. When densities of soil were 1.000, soil voids were − 1.000, while tallness, leaf count, flowering and okra output stood at −0.5249, −0.4904, −0.6115, −0.6825 and − 0.6909, respectively.

4.10.4 Regression of soil properties and growth parameters with yield of okra

Odey [25] reported an analytical work carried out on properties of soil—densification of soil and voids, development data and okra output, with the dependent variable as okra output (see results in **Tables 5** and **6**). Mathematical expression for okra output per hectare was determined from the analyzed results presented. Hence, predetermined model for okra output on agricultural field experiment conducted by different soil manipulations was inferred.

$$Y_o = -84.64 + 12.23_{\text{bd}} + 93.96_{\text{p}} + 0.21_{\text{H}} - 0.24_{NL} + 0.20_{\text{T}} \tag{5}$$

Where,
Y_o = okra output
Bd = soil bulk density
P = soil porosity
H = okra plant height
NL = numbers of leaves of okra
T = experimental treatment

The model eq. 5 above had a R^2 value, 0.934. The distinctions among mean values of estimated and observed okra output at 95% confidence level were applied to infer when statistical tools were used. According to Odey [25], t-test results revealed—no important difference at ($p > 0.05$) between experimented and predicted output of okra recorded when various soil manipulation methods were carried out. It can be

Parameters	Unstandardized coefficients		Standardized coefficients	T	Sig.	95.0% confidence interval for Beta	
	Beta	Std. error	Beta			Lower bound	Upper bound
(Constant)	-84.64	124.391		−.680	.545	−480.511	311.222
Bulk density	12.229	47.091	.206	.260	.812	−137.635	162.092
Porosity	93.963	124.816	.623	.753	.506	−303.257	491.184
Height	.212	.050	.906	4.264	.024	.054	.370
No leaves	−.244	.215	−.271	−1.137	.338	−.928	.440
Treatment	.201	.483	.070	.416	.705	−1.337	1.739

Reliant variable: Okra output Odey [25].

Table 5.
Regression coefficients.

Model	R	R square	Adjusted R square	Std. error of the estimate	Durbin-Watson
1	.966[a]	.934	.823	1.05255	1.165

aPredictors: (Constant), Treatment, No. of leaves, Bulk density, Height, Porosity.bDependent variable: Yield of okra Odey [25].

Table 6.
Model summary.

inferred that the forecasted output expression has close relation to the actual okra output. Therefore, the model is utilized in forecasting the production of okra in advance for any given agricultural field, provided other data are known. These findings agreed with the assertion of Mulumba and Lal [93], which stated that tillage practice is a management input that influences soil physical characteristics, which in turn touches growth and yield of crops.

Odey [25] concluded that the field study conducted was aimed at modeling okra growth and yield using different tillage systems, with the intention of predicting the output of okra. Field data were obtained and analyzed using appropriate tools. Findings revealed the negative relationship among development data, okra output and soil densities. Whereas, porosity of soil showed positive relationship among development criteria and okra output. Okra output was more in conventional and conservative tillage than in no till system. Modeled equation with R squared, 0.934 on the okra output using different tillage practices was created from the regression analysis, showing forecasted output almost equaled observed okra yield. The output of okra is estimated during cultivation, provided soil physical properties and growth parameters of okra are identified as in Odey et al. [94]. Odey [25] recommended conventional and conservative tillage systems for enhancement in the production of okra on soil with sandy loam texture.

5. Conclusion

Tillage, an integral component of crop production systems, exists with agricultural practice. It affects crop growth and yield. Different field experiments conducted by researchers on the effect of different tillage systems for predicting growth and yield of crops have been discussed. It has been revealed that estimating plant yield has comprehensive inferences for economics, ecology and human welfare. Moreover, a number of factors determining crop productivity make modeling crop production a challenging task. Forecasting crop production is more challenging, requiring making inferences on future performance based on past conditions. Farmers always try to focus on yield as soon as cropping activities commence. Finite model equations for predicting crop growth and yield are better than using farmer's previous experience on a particular crop. The chapter discussed how different tillage systems including conventional, conservative, reduced, zero tillage or direct seeding, mulch-tillage, ridge-tillage, stale seedbed, minimum tillage and strip-tillage are used in predicting crop growths and yields. Notable models for predicting growth and yield of crops using different tillage systems were highlighted. It is recommended that modern farming should utilize model equations for estimating crop growth and yield using soil physical properties, climate conditions and relevant parameters of appropriate tillage systems for improvement in the production of crops depending on soil types.

Author details

Simon Ogbeche Odey[1*] and Raphael Segun Bello[2]

1 Department of Wood Products Engineering, Cross River University of Technology, Calabar, Nigeria

2 Department of Agricultural and Bio-Environmental Engineering, Federal College of Agriculture, Ebonyi State, Nigeria

*Address all correspondence to: simonodey@unicross.edu.ng

References

[1] Mitchell JP, Pettygrove GS, Upadhyaya S, Shrestha A, Fry R, Roy R, et al. Classification of Conservation Tillage Practices in California Irrigated Row Crop Systems. Division of Agriculture and Natural Resources, University of California; 2009. DOI: 10.3733/ucanr.8364. Available from: http://anrcatalog.ucdavis.edu

[2] Upadhyaya SK, Lancas KP, Santos-Filho AG, Raghuwanshi NS. One-pass tillage equipment outstrips conventional tillage method. California Agriculture. 2001;**55**(5):44-47

[3] Madden NM, Southard RJ, Mitchell JP. Conservation tillage reduces PM10 emissions in dairy forage rotations. Atmospheric Environment. 2008;**42**:3795-3808

[4] Reicosky DC, Allmaras RR. Advances in tillage research in north American cropping systems. In: Shrestha A, editor. Cropping Systems: Trends and Advances. New York: Haworth Press; 2003. pp. 75-125

[5] Odey SO, Manuwa SI. Design steps of narrow tillage tools for draught reduction and increased soil disruption – A review. Agricultural Engineering International: CIGR Journal. 2016;**18**(1): 91-102

[6] Aikin SHM, Afuakwa JJ. Effect of four different tillage practices on cowpea performance. World Journal of Agricultural Science. 2010;**6**(6):644-651

[7] Arman K. Effect of different tillage system on soil properties and wheat yield in middle Anatolia. Soil and Tillage Research. 1997;**40**(3–4):201-207

[8] Ozpinar S, Clay A. Effect of tillage systems on the quality and crop productivity of a clayloam soil in semi-arid North-Western Turkey. Soil and Tillage Research. 2006;**88**(1–2):95-106

[9] Rashidi M, Keshavarzpour F. Effect of tillage methods on grain yield and yield components of maize (zee mays L.). International Journal of Agriculture and Biology. 2009;**9**(2):274-277

[10] Afolayan SO, Oladapo MO, Olaosebikan FO, Muibi EF. Effects of tillage systems and four fertilizer rates on growth parameters and fruit yield of okra (*Abelmoscus esculentus*) and pepper (*Capsicum annum*). World Journal of Agricultural Sciences. 2010;**6**(1):23-28

[11] Odey SO, Manuwa SI, Ademosun OC. Growth and yield response of soybean (ver TGX-1448-2E) to tractor traffic induced soil compaction in Sandy Clay soil of Akure, Nigeria. Nigerian Journal of Soil Tillage Research. 2014;**1**:62-74 ISSN 647 – 9921

[12] Odey SO. Overview of engineering problems of soil compaction and their effects on growth and yields of crops. European Journal of Advances in Engineering and Technology. 2018a; **5**(9):701-709

[13] Adekiya AO, Agbede TM, Aboyeji CM, Adegbite KA, Dunsin O, Adekanye TA, et al. Soil properties, okra performance and nutrient compositions as affected by tillage and maize cob ash. Journal of Crop Science and Biotechnology. 2019;**22**(2):113-122

[14] Ozpinar S, Isik A. Effect of tillage ridging and row spacing on seedling emergence and yield of cotton. Soil and Tillage Research. 2004;**75**(1):19-26

[15] Allmaras RR, Dowdy RH. Conservation tillage system and their

adoption in the United States. Soil Tillage Research. 1985;**5**:197-222

[16] Ruberson, J. R. And S. C. Phatak. 1997. Insect population in a cover crops/strip tillage system. Proceedings of Belt Wide Cotton conference. Vol. 2 pp. 1121-1124.

[17] Mamkagh AMA. Effect of tillage time and plastic mulch on growth and yield of okra (*Abelmoschus esculentus*) grown under rainfed conditions. International Journal of Agriculture and Biology. 2009;**11**(4):453-457

[18] Olaoye JO, Oyeleke IF, Olaoye MO. Effects of tillage operations on the yield components of okra. In: A Presentation at the International Soil Tillage Organization (ISTRO, 2018), Nigeria: Conference Proceedings of International Soil Tillage Research Organization (ISTRO, 2018); 2018

[19] Rasmussen KJ. Impact of ploughless soil tillage on yield and quality: A Scandinavian review. Soil and Tillage Research. 1999;**53**(1):3-14

[20] MWPS (MidWest Plan Service). Conservation tillage systems and management. In: Crop Residue Management with No-Till, Ridge-Till, Mulch-Till and Strip-Till. MWPS-45. 2nd ed. Ames: Iowa State University, MidWest Plan Service; 2000 MWPS Web site, http://www.mwps.org/index.cfm?fuseaction=c_Products.viewProduct&cat_ID=720&productID=6482&skunumber=MWPS-45&crow=4

[21] Owen H. Tillage: From Plow to Chisel and no Tillage 1930–1999. Ames Lowa State University Midwest Plant Services; 2001. DOI: 10.3390/su12031062

[22] ASAE (American Society of Agriculture and Biological Engineers). Terminologies and Definitions for Soil Tillage and Soil Tools Relationship ASEAE EP291. Feb 2005. pp. 131-134

[23] Leonard BR, Torey K, Hutchinson RL. Influence of conservation tillage on cotton insect pest ecology: A case study with cotton aphid (aphis gossypii glover). In: Proc. 23rd Annual Southern Conservation Tillage Conference for Sustainable Agriculture. USA: North little Rock Ar; 2000. pp. 43-44

[24] Hedayatipoor A, Alamooti MY. Effect of conservative tillage on physical properties of soil and yield of rainfed wheat. Agricultural Engineering International: CIGR Journal. 2020;**22**(1):48-53

[25] Odey SO. Finite modeling of growth and yield of okra using different tillage systems. Agricultural Engineering International: CIGR Journal. 2022;**24**(1):188-196

[26] Agbede TM. Effect of tillage and oil palm bunch ash plus poultry manure on soil chemical properties, growth and ginger yield. World Academy of Science, Engineering and Technology. International Journal of Agricultural and Biological Engineering. 2013;**7**(7):481-488

[27] Magdoff O, Van Es H. Building Soils for Better Crops. 2nd ed. Jarboe Printing Washington D. C. 240P: A Sustainable Agriculture Networking Publication; 2000

[28] Soane BD, Ouwerkerk VC. Soil compaction in crop production. In: Developments in Agricultural Engineering Series. Vol. 2. Amsterdam: Elsevier; 1994. pp. 1-121

[29] Manor G, Clark RL. Development of an instrumented Subsoiler to map soil

hard-pans and real-time control of Subsoiler depth. In: ASAE Annual International Meeting, Sacramento Convention Center, Transactions of the ASAE (American Society of Agricultural Engineers), Sacramento, California, USA, August, 2001. 2001

[30] Petersen M, Ayers P, Westfall D. Managing soil compaction. CSU Cooperative Extension Agriculture. 2004;**0**:519

[31] Wells LG, Stombaugh TS, Shearer SA. Crop yield response to precision deep tillage. Transactions of the ASAE. 2005;**48**(3):895-901. American Society of Agricultural Engineers ISSN 0001–2351

[32] Alameda D, Villar R. Moderate soil compaction: Implications and growth and architecture in seedlings of 17 woody plant species. Soil and Tillage Research. 2009;**103**:325-331

[33] Mason EG, Cullen AWJ, Rijkse WC. Growth of two Pinus radiata stock types on ripped and ripped/bedded plots at Karioi Forest. New Zealand Journal of Forestry Science. 1988;**18**: 287-296

[34] Aase JK, Bjomeberg DI, Sojka RE. Zone subsoiling relationships to bulk density and cone index on a furrow-irrigated soil. Transactions of ASAE. 2001;**44**:577-583

[35] Raper RL, Reeves DW, Burt EC. Using in-row subsoiling to minimize soil compaction caused by traffic. Journal of Cotton Science. 1998;**2**(3):130-135

[36] Taylor JH. Benefits of permanent traffic lanes in a controlled traffic crop production system. Soil Tillage Research. 1983;**3**:385-395

[37] Monroe CD, Kladivko EJ. Aggregate stability of a silt loam as affected roots of maize, soybean and wheat. Communications in Soil Science and Plant Analysis. 1987;**18**:1077-1087

[38] Mari GR, Changying J. Influence of agricultural machinery traffic on soil compaction patterns, root development, and plant growth, overview. American-Eurasian Journal of Agric. & Environmental Sciences. 2008;**3**(1): 49-62 ISSN 1818-6769.

[39] Tardieu F. Growth and functioning of roots and of root systems subjected to soil compaction. Towards a system with multiple signaling. Soil Tillage Research. 1994;**30**:217-243

[40] Dauda A, Samari A. Cowpea yield response to soil compaction under tractor traffic on a sandy loam soil in the semi-arid region of northern Nigeria. Soil Tillage Research. 2002;**68**:17-22

[41] Amauri NB, Jasse FC, Maria AP, Silva FO, Eurica LS, Cristian LS, et al. Traffic soil compaction of oxisol related to soybean development and yield. Journal of Agriculture Science (Piracicaba, Brazil),. 2007;**64**(6): 608-615

[42] Isaac NE, Talor RK, Staggenborg SA, Schrock MD, Leikam DF. Using cone index data to explain yield variation within a field. Agricultural Engineering International: the CIGR Journal of Scientific Research and Development. 2002;**IV**:1-13

[43] Borghei AM, Taghinejad J, Minaei S, Karimi M, Varnamkhasti MG. Effect of subsoiling on soil bulk density, penetration resistance and cotton yield in northwest of Iran. Int. J. Agri. Biol. 2008;**10**:120-123

[44] Weber R, Biskupski A. Effect of penetration resistance, bulk density and moisture content of soil on selected yield

components of winter triticale crop yield in relation to method of cultivation. International Agrophysics. 2008;**22**: 171-177

[45] Soltanabadi MH, Miranzadeh LM, Karimi LM, Varnamkhasti MG, Hemmat A. Effect of subsoiling on soil physical properties and sun flower yield under conditions of conventional tillage. International Agrophysics. 2008;**22**: 313-317

[46] Grzesiak MT. Impact of Soil Compaction on Root Architecture, Leaf Water Status, Gas Exchange and Growth of Maize and Triticale Seedlings. Niezapominajek 21, PL 30-239./Plant Root 3: The Franciszek Górski Institute of Plant Physiology, Polish Academy of Sciences; 2009. pp. 10-16. DOI: 10.3117/plantroot.3.10

[47] Juliano CC, Rosolem CA. Soybean root growth and yield In rotation with cover crops under Chiseling and No-till. European Journal of Agronomy. 2010; **33**(3):242-249

[48] Kulkarni SS, Bajwa SG, Huitink G. Investigation of the effects of soil compaction in cotton. Transactions of the ASABE. ISSN 2151–0032. 2010;**53**(3): 667-674

[49] Becerra AT, Botta GF, Lastra Bravo X, Tourn M, Bellora Melcon F, Vazquez J, et al. Soil compaction distribution under tractor traffic in almond (*Prunus amigdalus* L.) orchard in Almeria Espana. Soil & Tillage Research. 2010;**107**:49-56

[50] Becerra AT, Tourn M, Botta GF, Lastra Bravo X. Effects of different tillage regimes on soil compaction, maize (Zea mays L.) seedling emergence and yields in eastern Argentina pampas region. Soil & Tillage Research. 2011;**117**: 184-190

[51] Chen G, Weil RR. Root growth and yield of maize as affected by soil compaction and cover crops. Soil and Tillage Research. 2011;**117**:17-27

[52] Celik A, Raper RL. Design and evaluation of ground-driven rotary subsoilers. Soil and Tillage Research. 2012;**124**:203-210

[53] Li X, Zhang D, Zhang R, Osman AN. Performance of an oscillating subsoiler in reducing resistance. In: Presentation of the American Society of Agricultural and Biological Engineers. Paper No. 12–1341191. St. Joseph, Michigan; 2012

[54] Al-Adawi SS, Reeder RC. Compaction and subsoiling effects on corn and soybean yields and soil physical properties. Transactions of the ASAE. 1996;**39**(5):1641-1649

[55] Reeder RC, Wood RK, Finck CL. Five Subsoiler designs and their effects on soil properties and crop yields. Transactions of the American Society of Agricultural Engineers. 1993;**36**(6): 1525-1531

[56] Abu-Hamdeh NH. Compaction and subsoiling effects on corn growth and soil bulk density. Soil Science Society of America Journal. 2003;**67**(4):1213-1219

[57] Lipiec L, Medvedev VV, Birkas M, Dumitru E, Lyndina TE, Rousseva S, et al. Effect of soil compaction on root growth and crop yield in central and Eastern Europe. International Agrophysics. 2003;**2003**(17):61-69

[58] Botta GF, Tolon Becerra A, Lastra Bravo X, Tourn M. Tillage and traffic effects (planters and tractors) on soil compaction and soybean (Glycine max L.) yields in Argentinean pampas. Soil & Tillage Research. 2010;**110**: 167-174

[59] Nolting SP, Edwards CR. Yield response of soybean to defoliation by the Mexican bean beetle (Coleoptera: Coccinellidae). Journal of Economic Entomology. 1989;**82**:12121218

[60] Browde JA, Pedigo LP, Owen MDK, Tylka GL. Soybean yield and pest management as influenced by nematodes, herbicides, and defoliating insects. Agronomy Journal. 1994;**86**: 601-608

[61] Grymes CF, Griffin JL, Boethel DJ, Leonard BR, Jordan DL, Russin JS. Soybean response to weed interference and defoliation. Weed Science. 1999;**47**: 90-94

[62] Board JE, Harville BG. Explanations for greater light interception in narrow- vs. wide-row soybean. Crop Science. 1992;**32**:198-202

[63] Board JE, Wier AT, Boethel DJ. Critical light interception during seed filling for insecticide application and optimum soybean grain yield. Agronomy Journal. 1997;**89**:369-374

[64] Agba OA, Mbah BN, Asiegbu JE, Adinya IB. Effects of spacing on the growth and yield of okra (Abelmochus Esculentus) in Obubra, Cross River State. Global Journal of Agricultural Sciences. 2011;**10**(1):57-61

[65] Hunt R. Basic Growth Analysis. London: Unwin Hyman Ltd.; 1990. p. 112

[66] South DB. Relative growth rates: A critique. South African Forestry Journal. 1995;**173**:43-48

[67] Hoffmann WA, Poorter H. Avoiding bias in calculations of relative growth rate. Annals of Botany. 2002;**90**(1):37

[68] Briggs WL, Cochran L, Gillett B. Calculus: Early Transcendentals. Pearson Education, Limited; 2011. p. 441. ISBN 978-0-321-57056-7. Available from: https://books.google.com.ng/books/about/Calculus.html?id =_cMLQg AACAAJ&redir_esc=y [Retrieved: September 24, 2012]

[69] Paine CET, Marthews TR, Vogt DR, Purves D, Rees M, Hector A, et al. How to fit nonlinear plant growth models and calculate growth rates: An update for ecologists. Methods in Ecology and Evolution. 2012;**3**(2):245

[70] Fredericksen TS, Zedacker SM, Seiler JR. Interference interaction in simulated pine-hardwood seedling stands. Forest Science. 1993;**39**:383-395

[71] Margolis HA, Brand DJ. An ecological basis for understanding plantation establishment. Canadian Journal of Forest Research. 1990;**20**:375-390

[72] Osonubi O, Osundina MA. Comparison of the response of flooding of seedlings and cuttings of *Gmelina*. Tree Physiology. 1987;**3**:147-156

[73] Samuelson LJ, Seiler JR. Interactive role of elevated CO_2, nutrient limitations, and water stress in the growth responses of red spruse seedlings. Forest Science. 1993;**39**:348-358

[74] Manjula E, Djodiltachoumy. A model for prediction of crop yield. International Journal of Computational Intelligence and Informatics. 2017;**6**(4):298-305

[75] Sangeeta SG. Design and implementation of crop yield prediction model In agriculture. International Journal of Scientific & Technology Research. 2020;**8**(01):544-549

[76] Odey SO. Tractor traffic influence on soil properties, growth and yield of maize in Obubra, Nigeria. International

Journal of Engineering Inventions. 2018b, 2018;**7**(01):01-10

[77] Nkakini SO, Davies RM. Mathematical model equation for tolerance of okra plant yield to soil densification. International Journal of Research and Innovation in Applied Science (IJRIAS) |. 2020;**V**(I):80-92 ISSN 2454-6194.

[78] Reddy KR, Reddy V, Hodges HF. Effect of temperature on early season cotton growth and development. Agronomy Journal. 1992;**84**:229-237

[79] Omran AF, Barkry AI, Gawish AR. Effect of soaking seeds in some growth regulator solutions on the growth, chemical constituent and yield of okra, seed Sci. Technology. 1980;**8**(1): 161-168

[80] Khan VA, Stevens C, Brown JE. Early response of transplanted Okra grown under VisPore row cover and two types of polyethylene mulch. In: Dube PA, Stewart K, editors. Proceedings of the 22nd National Agricultural Plastics Congress. Madison: University of Wisconsin; 1990. pp. 28-32

[81] Black GR, Hartge KH. Bulk density. In: Klute A, editor. Methods of Soil Analysis. Part I. Physical and Mineralogical Methods. 2nd. ed. Madison, Wisconsin, USA: Agronomy No. 9 (part I). ASASSSA; 1986. pp. 363-375

[82] Gee GW, Bauder JW. Particle size analysis. In: Klute A, editor. Method of Soil Analysis, Part 1-Physical and Mineralogical Methods. Madison, Wiscosin, USA: SSSA Book Serial 5; 1986. pp. 383-412

[83] Udo EJTO, Ibia JA, Ogunwale AO, Esu IE. Manual of Soil, Plant and Water Analysis. Lagos: Sibon Books Limited; 2009

[84] Nelson DM, Sommers LE. Total carbon, organic carbon and organic matter. In: Spark DL, editor. Methods of Soil Analysis Pact 3-Chemical Methods. Madison, Wisconsin USA: SSSA Book series 5; 1996. pp. 961-1010

[85] Kuo S. Phosphorus. In: Sparks DL, editor. Methods of Soil Analysis: Part 3. Madison: SSSA Book Series No. 5, SSSA and ASA; 1996. pp. 869-919

[86] Bremmer JM. Total nitrogen. In: Spark DL, editor. Methods of Soil Analysis Pact 3-Chemical Methods. Madison, Wisconsin, USA: SSSA Book Series 5; 1996. pp. 1085-1111

[87] Sumner ME, Miller WP. Cation exchange capacity and exchange coefficients. In: Sparks DL, editor. Methods of Soil Analysis Part 3: Chemical Methods, SSSA Book Series 5. Madison, Wisconsin: Soil Science Society of America; 1996. pp. 1201-1230

[88] Ojeniyi S. Tropical Soil Management and Agriculture in Nigeria. Ibadan Nigeria: Bash-Moses Publishers; 2010

[89] Brady N, Weil R. The Nature and Properties of Soil. 13th ed. Upper saddle River New Jersey: Prentice Hall; 2002. p. 960P

[90] Carter MR. Soil quality for sustainable land management: Organic matter and aggregation interactions that maintains soil functions. Agronomy Journal. 2002;**94**:38-47

[91] Lal R, editor. Encyclopaedia of Soil Science. Boca Raton FL: Taylor and Francis; 1991. pp. 1333-1335

[92] Ariyo OJ, Akenova CA. Plant Character and Correlation and Path Analysis of Pod Yield in Okra. PMB 5320 Ibadan: International Institute of Tropical Agriculture; 1986

[93] Mulumba LA, Lal R. Mulching effect on selected soil physical properties. Soil and Tillage Research. 2008;**98**(1):106-111

[94] Odey SO, Manuwa SI, Ewetumo T. Development and performance evaluation of instrumented Subsoilers in breaking soil hard-pan. Agricultural Engineering International: CIGR Journal. 2018;**20**(3):85-96

Chapter 10

Harnessing Soil Potential: Innovation in Strategic Tillage and Management – New Perspectives

Sajid Ali, Adnan Zahid, Ammara Fatima, Mukhtar Ahmad, Muhammad Tariq Manzoor, Asma Ayub, Ahmad Raza and Nauman Shafqat

Abstract

The sustainability of the environment and the productivity of agriculture are both critically dependent on soil. Maximizing agricultural yields while reducing agriculture's negative environmental effects is becoming more and more important as the world's population continues to expand. Innovating tillage and management techniques to harness the potential of the soil is a topic that is explored in this chapter. The first section of the chapter describes the difficulties that contemporary agriculture faces, such as soil erosion, nutrient depletion, and water shortages. The part new technology has played in managing soil. Making educated management decisions is made easier by using precision agricultural technology like soil sensors, remote sensing, and geographic information systems (GIS). These technologies provide useful insights into soil variability. It emphasizes how crucial it is to implement sustainable soil management techniques in order to guarantee long-term agricultural output and ecological harmony. The chapter's conclusion emphasizes the need of maximizing soil potential through creative methods of tactical tillage and management. Agricultural systems may raise crop yield, lessen their environmental effect, and become more resilient to climate change by using sustainable soil practises, assuring a more sustainable and food-secure future.

Keywords: sustainability, innovative tillage, nutrient depletion, remote sensing, soil potential

1. Introduction

Degraded soil and less ecosystem services, which in turn promotes unsustainable agricultural development [1]. This is true even though intensive agriculture is crucial for meeting the world's growing food demand [2]. One of the most important anthropogenic activities that have had a substantial impact on soil health is the intensive use of farms and tillage practices. A key issue affecting the sustainability and productivity of the soil is the degradation of the soil's health brought on by multiple-cropping systems and tillage. Tillage intensive cropping practices have deteriorated soil chemical, physical, and biological characteristics and decreased soil organic carbon (SOC) levels [3]. Soil ecosystems to function and perform sustainably, it is essential to safeguard and improve soil function and health [4, 5]. It is possible to establish whether diverse soil management strategies have the necessary effectiveness for soil functions and productivity if multiple soils attributes and processes are merged into a single estimate value of soil health [6, 7]. The effect of the tillage system on soil characteristics must be taken into account in order to maintain soil fertility and assess the sustainability of agricultural systems [8–10]. By using a system of soil minimal tillage, the amount of larger vegetal leftovers (minimum 30%) that are left at the soil surface and in the upper 10–20 cm layers of the soil, in various phases of decomposition, shows an increasing tendency. Through the use of minimal systems, humus determination after 4 years shows an increasing trend, with an increase of up to about 0.41%. Conservative agriculture impacted on organic carbon accumulates in soil at a rate between 0.27 and 1.10 tha^{-1} $year^{-1}$ [11–15] which ultimately improve the soil health. Strategic tillage is a practice in agriculture that involves the targeted and purposeful manipulation of soil through mechanical means to achieve specific objectives. It encompasses various techniques used for soil preparation before planting and for managing soil conditions after planting. Tillage tools are employed to bring about desired effects such as pulverization, cutting, or movement of the soil. The primary goals of strategic tillage are to enhance soil structure, control weeds, manage crop residues, improve water intake and storage, facilitate root development, and promote optimal growing conditions for crops [16].

2. Fundamentals of strategic tillage

Strategic tillage involves different types of tillage operations, including primary tillage and secondary tillage. Primary tillage is performed to break and loosen the soil for a significant depth, typically ranging from 15 to 90 cm (6–36 inches). It includes various equipment such as moldboard plows, disk plows, rotary plows, chisel plows, and subsoil plows [17]. These tools are used based on the soil type, the desired depth of tillage, and the specific objectives of the tillage operation.

2.1 Key principles and objectives

The key principles and objectives of strategic tillage can be summarized as follows: (i) *Soil structure modification*: strategic tillage aims to improve soil structure by creating desirable conditions for optimal plant growth. This helps in creating an environment conducive to root development and moisture retention [18]. Strategic tillage, according to research, can drastically affect soil structure and porosity, resulting in better water transport and root growth [19, 20]. It can contribute to the long-term sustainability and productivity of agricultural systems [21]. Research studies have indicated that strategic tillage can improve soil biological activity, increase nutrient availability, and enhance soil organic carbon content [22]. (ii) *Weed control*: tillage plays a crucial role in weed management by disrupting weed growth, burying weed seeds, and uprooting established weeds [16]. Studies have demonstrated the effectiveness of strategic tillage in suppressing weed populations and reducing weed biomass [23, 24]. Tillage plays a crucial role in controlling weeds. Strategic tillage practices, such as timely tillage operations and burying weed seeds, can help reduce weed populations and minimize the need for herbicides [25]. (iii) *Crop residue management*: strategic tillage helps incorporate agricultural byproducts in the ground, facilitating their decomposition and nutrient release [26]. Strategic tillage facilitates the incorporation of crop residues and soil modifications such as lime and manure, into the soil. This helps distribute nutrients more evenly, improves nutrient availability to plants, and reduces nutrient losses through runoff [27]. Strategic farming, according to research, can improve the breakdown of crop residues, leading to improved nutrient cycling and soil organic matter content [28]. (iv) *Water intake and storage*: strategic tillage optimizes the uptake, storage, and transfer of water in the soil profile by enhancing soil structure. It promotes better water infiltration, reduces runoff, and increases water-holding capacity, thus mitigating the risks of drought stress and improving crop water-use efficiency [29]. (v) *Root zone deepening*: strategic tillage can help break through compacted layers, such as hardpans, and enhance root penetration into the subsoil. This enables crops to access deeper water reserves and nutrients, leading to improved yields and plant health [30]. (vi) *Integrated weed and pest management*: strategic tillage can be integrated into broader weed and pest management strategies. By disrupting weed life cycles and providing favorable conditions for natural predators, tillage can contribute to reducing the reliance on herbicides and pesticides [31]. Studies have highlighted the potential of strategic tillage in integrated weed management approaches, leading to reduced herbicide use and improved weed control [32].

3. Recent innovations in tillage equipment

3.1 Modern tillage equipment

Modern tillage equipment has seen several advancements in recent years, aiming to improve efficiency, soil health, and sustainability in agricultural practices. Some notable innovations include:

(i) *Strip-till rigs*: strip-tillage is gaining popularity as a conservation-minded practice. Strip-till rigs are designed to till narrow strips, typically 6–12 inches wide, between rows. This approach reduces soil erosion, speeds up soil warming, conserves energy and fuel, and maintains higher levels of soil organic matter [33]. (ii) *Vertical-tillage implements*: with vertical tillage, crop residue is broken down and the top 2–3 inches of soil are loosen. It improves soil contact with residue, enhances infiltration, reduces runoff, and creates more uniform field conditions. The advantages of traditional disks, vertical tillage, and soil-finishing products have been combined into a single device by equipment manufacturers [34]. (iii) *Rotary finishers*: rotary finishers are innovative tools used for soil conditioning. They incorporate rotary harrows or rolling baskets to further refine the soil surface after primary tillage operations. These implements help improve seedbed preparation, promote soil structure, and enhance moisture retention [35]. (iv) *Soil conditioners*: soil conditioners are specialized machines used to enhance soil quality. They can incorporate organic matter into the soil, break up compacted layers, and improve soil structure. Soil conditioners help promote root growth, increase nutrient availability, and optimize water infiltration [36].

3.2 Precision tillage tools and implements

Precision tillage focuses on precise control of tillage operations to optimize seedbed conditions, minimize soil disturbance, and preserve soil structure. Some recent innovations in precision tillage tools and implements include: (a) *TruSet technology*: advanced precision control systems like TruSet technology allow farmers to adjust shank depth in small increments (e.g., 0.10 inch) from the tractor cab. These systems offer precise depth control and can optimize tillage operations based on soil conditions and crop requirements [37]. (b) *Reduced-till implements*: reduced-till implements, such as chisel plows and disk rippers, have gained popularity due to their ability to perform efficient tillage while minimizing soil disturbance. These implements allow farmers to maintain residue cover, reduce erosion, and conserve soil moisture [38].

3.3 Conservation tillage machinery

Conservation tillage focuses on minimizing soil disturbance and maintaining residue cover to protect soil health and reduce erosion. Recent innovations in conservation tillage machinery include:

(i) *No-till seeders/drills*: no-till seeders or drills are designed to plant seeds directly into untilled soil, leaving crop residue undisturbed. These machines have precise seed placement mechanisms that help ensure optimal seed-to-soil contact while preserving soil structure and minimizing erosion risks [39]. (ii) *Cover crop inter-seeding equipment*: inter-Seeding cover crops between cash crop rows is an effective conservation practice. Specialized equipment has been developed to accurately seed cover crops in standing cash crops without causing significant disturbance. This approach enhances soil health, reduces weed pressure, and improves nutrient cycling [40].

3.4 No-till and reduced-till systems

No-till and reduced-till systems have gained traction as sustainable alternatives to intensive tillage. Recent developments in these systems include: *No-till planters*: no-till planters are equipped with features that enable precise seed placement into untilled soil. They typically incorporate residue managers, row cleaners, and depth control mechanisms to facilitate successful planting while preserving soil structure and moisture [41]. *Conservation-ready tractors*: manufacturers have introduced tractors equipped with specialized features for conservation tillage, such as advanced guidance systems, reduced compaction tire technology, and variable-rate planting capabilities. These tractors offer improved efficiency and precision in implementing no-till and reduced-till practices [42].

4. Soil health improvement techniques

Utilizing cover crops is one method for enhancing soil health and environmental quality [43]. When the primary cash crops are not growing, certain crops, referred to as cover crops, are planted instead. These cover crops were chosen because of their range of beneficial effects on the soil.

Benefits of cover cropping for soil health improvement: 1. *Erosion control*: cover crops help to minimize soil erosion by providing ground cover and protecting the soil surface from wind and water erosion [44]. The risk of soil loss is decreased by the roots of cover crops holding the soil particles together. 2. *Nutrient management*: by scavenging and absorbing surplus nutrients from the soil, cover crops can stop these nutrients from evaporating into groundwater or being lost through runoff. Following the termination of the cover crops and their incorporation into the soil, the released nutrients become available for the succeeding cash crops [45]. 3. *Adding organic matter*: cover crops decompose and increase the amount of organic matter in the soil. Organic matter enhances the soil's structure, ability to retain water and nutrients, microbial activity, and overall health [46]. 4. *Controlling weed growth*: cover crops can inhibit the growth of weeds by competing with them for sunlight, nutrients, and space. Herbicides and manual weed management are not necessary because of the dense cover crop canopy's ability to shade out emerging weeds and these crops also have allopathic effects [47]. 5. *Biological diversity*: cover crops enhance soil microbiome [48]. Beneficial insects, earthworms, and other soil organisms can find a home and food source in cover crops. These species support healthy soil ecosystems, pest management, and nutrient cycling [49]. 6. *Controlling soil moisture*: some cover crops, like legumes, have deep roots that can improve the soil's structure and drainage. By lowering evaporation from the soil surface, they also assist in conserving soil moisture [50]. Considerations for cover crop implementation should include the regional climate, the kind of soil, and any special goals for soil development [51]. Different cover crop varieties and species can be adapted to meet certain requirements for soil health and work with crop rotation systems [52].

5. Soil amendment and fertilization practices

By promoting soil fertility, nutrient availability, and general soil structure, soil amendment and fertilization practices are essential for enhancing soil health. The following are some essential components of fertilization and soil amendment techniques and how they help to promote soil health: *organic matter addition*: rich sources of organic matter include crop wastes, compost, and manure, improved soil structure, moisture retention, and nutrient-holding capacity [53]. Additionally, feeding beneficial soil organisms, organic matter encourages microbial diversity and activity. *Refueling with nutrients*: fertilizers are used to replenish vital nutrients that the soil may be lacking. In addition to secondary macronutrients like calcium, magnesium, and sulfur, as well as micronutrients like iron, zinc, and manganese, among others, the primary macronutrients for plant growth are nitrogen (N), phosphorus (P), and potassium (K) [54]. Crops are guaranteed to acquire enough nutrients for the best development and production when fertilization is properly balanced. *Nutrient cycling and recycling*: cycling and recycling of nutrients in the soil ecosystem are made possible by organic fertilizers and soil amendments. As nutrients are released through the breakdown of organic materials, plants may now absorb them [55]. By transforming and moving nutrients through biological, chemical, and physical processes, nutrient cycling helps to increase nutrient usage efficiency while minimizing nutrient losses. *pH adjustment*: soil amendments can assist with pH adjustment, which impacts the availability of nutrients. In alkaline soils, elemental sulfur or acidic organic materials can be used to reduce soil pH, whereas lime is frequently used to raise soil pH in acidic soils [56]. The availability and uptake of nutrients by plants are maximized when soil pH is kept at the right level [57]. *Microbial activity promotion*: organic fertilizers and soil additives give beneficial soil microorganisms a home and a food source. These microorganisms, which include bacteria, fungi, and earthworms, support the Composting of organic materials, nutrient cycling, and the prevention of disease. Improved soil structure, nutrient availability, and general soil health are all benefits of healthy microbial communities [58]. *Sustainable nutrient management*: this seeks to lessen nutrient losses that could harm ecosystems and water quality, such as leaching or runoff. Improved fertilizer input rates, timing, and locations can be achieved by split applications, precision farming, and soil testing, all of which reduce adverse environmental consequences and increase soil health. It is crucial to remember that fertilization and soil amendment techniques should be adjusted to the particular requirements and characteristics of the soil and the crops being grown. Effective and long-term maintenance of soil fertility requires soil testing, nutrient management strategies, and adherence to recommended application rates. Additionally, adopting practices like organic farming, crop rotation, and cover crops can enhance soil health and long-term sustainability by working in conjunction with soil amendment and fertilization techniques.

6. Precision agriculture and soil mapping

Precision farming, often referred to as agricultural precision or smart farming, is a cutting-edge method of managing agriculture that makes use of technology to enhance production, efficiency, and sustainability in farming practices [59]. It entails utilizing a variety of technologies to gather and analyze data in order to give farmers all the knowledge they need to manage their crops and livestock. Several crucial technologies for accurate agriculture include:

1. Global positioning system (GPS) technology: GPS enables precise mapping and tracking of agricultural machinery and field operations [60]. This aids farmers in maximizing planting, irrigation, and harvesting, ensuring that each process is carried out precisely and with the least amount of overlap possible.

2. Geographic information system (GIS): GIS creates precise maps of farmers' fields by fusing information from GPS with additional geographic information. They can use this information to recognize variations in soil types, moisture levels, and other characteristics, allowing for management techniques that are site-specific.

3. Real-time information on crop health, water stress, and insect infestations can be obtained via remote sensing devices [61], like satellites and drones. Producers can use this knowledge to spot problems early and execute targeted treatments.

4. Soil sensors are used in the field to detect vital characteristics such as soil moisture, nutrient levels, pH, and others. With the aid of this information, farmers can accurately apply fertilizer and irrigation, minimizing waste and maximizing crop development.

5. Weather monitoring: farmers can make educated decisions about planting, harvesting, and other significant operations by using weather stations on the farm or having an understanding of local meteorological data [62].

6. With the help of a technology known as variable rate technology (VRT), farmers can adjust the rates at which inputs like insecticides are used and fertilizers are applied based on the particular requirements of various fields [63]. This limits abuse while reducing environmental effects and increasing production.

7. Automated machinery: autonomous or semi-autonomous machinery can carry out operations with great precision, such as planting, spraying, and harvesting, which lowers labor expenses.

8. Internet of Things (IoT) devices, including sensors and actuators, are utilized to remotely control and automate a range of farming operations while giving real-time data [64] to their scalable and environmentally friendly capabilities [65]. These tools deliver insightful infoPrecision farming, often referred to as agricultural precision or smart farming, is a cutting-edge method of managing agriculture that makes use of technology to enhance production, efficiency, and sustainability in farming practices [59]. It entails utilizing a variety of technologies to gather and analyze data in order to give farmers all the knowledge they need to manage their crops and livestock.

7. Soil erosion processes

Water erosion, wind erosion, tillage erosion, and soil loss during crop harvesting are the typical classifications of soil erosion (sometimes known as "harvest erosion"). The primary driver distinguishes the various erosion kinds, as indicated by the nomenclature. Water (such as precipitation and snowmelt) is the primary cause of water erosion, and its various forms include sheet erosion, rill erosion, and gully erosion [66]. Wind, the main driver of wind erosion, causes many types of soil transport, including creeping, saltation, and suspension, depending on the particle size [67]. The type of tillage tool being used (such as a mouldboard plow or chisel plow) affects how much dirt is moved throughout the plowing process [68]. Soil loss occurs during crop harvesting when crops are harvested and the harvested product comes into touch with the soil directly, as in the case of sugar beet and potatoes [69]. Although all types of soil erosion are significant environmental processes, the number of peer-reviewed papers is drastically out of proportion as shown in **Figure 1**. While water erosion

Figure 1.
Number of peer-reviewed publications on the subject of soil erosion (Web of Science and Scopus search, 17 February 2021) [70].

predominates in publications (10,505–16,817 research), agricultural harvesting-related soil loss is almost completely ignored (29–36 studies). This may be explained by the fact that, according to [71], water erosion accounts for 35.9 Petagrams of carbon (Pg) of soil loss annually, making it the most significant soil erosion process in the world.

One of the biggest problems in many countries today is soil erosion, which is a global concern. According to [72], the process called soil erosion through which soil is destroyed as a result of a combination of natural (such as water, wind, and snow) and man-made (such as intense and extensive agriculture) forces. In terms of the aquatic and terrestrial environments, the effects of soil erosion on agricultural output, source water quality, and ecosystem health are detrimental [73]. The primary causes of soil erosion include geography, ground cover, climate erosivity, and soil erodibility. Agriculture output is based on soil. Soil erosion in the agricultural area has put the sustainability of agricultural activities in jeopardy. Accelerated soil erosion harms both the environment and the economy [74]. Both on-site and off-site productivity may have been impacted by soil erosion. The loss in output from soil erosion that occurs both on-site and off-site is attributed to three interrelated consequences: a drop in soil quality, long-term productivity effects, and short-term productivity effects [75]. Two effects of soil erosion that render the area concerned unsuitable for agriculture and affect the productivity of agricultural land are denudation of topsoil and a loss in soil fertility. Asia is home to the majority of the world's tropical and subtropical fruit production, which accounts for 90% of the world's rice production [76]. Asia is also well-known for its vast plantations of key cash crops including tea, palm oil, coconut, sugarcane, and rubber. Sadly, agriculture activities have a negative impact on the environment, namely soil erosion, despite giving Asia a healthy supply of riches. Asia has possibly had the most severe soil erosion of any continent.

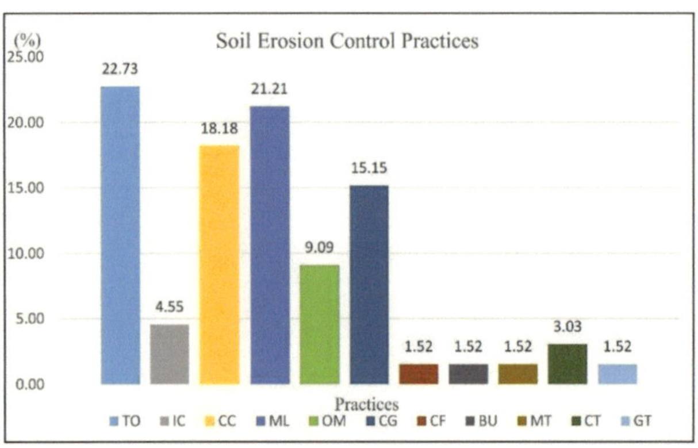

Agronomics practices		**Agrostological practices**	**Mechanical practices**	
TO=Tillage operation IC= Intercropping CC= Cover crop	ML=Mulching OM= Organic matter	CG= Cultivation of grass	CF= Contour farming BU= Bund MT=Micro basin tillage	CT= Contour terracing GT=Geo-textile

Figure 2.
Research on soil erosion control practices in Asia's agricultural land [83].

Numerous studies on soil erosion management techniques are carried out every year across Asia. Studies mostly concentrate on the impact of various management practices, such as tillage operation [77], mulching [78], cover crop [79], and intercropping [80] on runoff generation and erosion process. Studies on control methods have also been found to be extremely encouraging outside of Asia. Additionally, those studies use a variety of techniques, including straw mulch [81] and catch crop [82]. Researchers in the linked subject have consistently produced studies on how to address soil erosion issues and produced a number of sometimes-conflicting findings. These can be the result of sampling variation, study faults, or discrepancies in the studies. As a result, it is unclear if the most accurate outcomes should be employed or used in the real world. So, in order to locate previously conducted research, a systematic review is required.

According to **Figure 2**, over the previous 6 years, tillage operations (22.73%) have been tested the most frequently, followed by mulching (21.21%), cover crops (18.18%), grass culture (15.15%), and other management techniques.

8. Economic and environmental implications

1. *Cost-benefit analysis of strategic tillage*: cost-benefit analysis (CBA) is a systematic approach used to assess the pros and cons of a specific project or activity. In the context of strategic tillage, it involves evaluating the expenses associated with implementing strategic tillage practices and comparing them to the benefits it offers. Strategic tillage refers to targeted and controlled tillage techniques that aim to achieve predetermined goals while minimizing negative environmental impacts [84]. The costs of strategic tillage include equipment and machinery expenses, labor costs, training and education expenses, fuel and maintenance costs, and potential environmental impacts. On the other hand, the benefits include improved soil structure, better weed control, reduced soil erosion, enhanced water conservation, increased crop yields, and cost savings compared to traditional tillage methods [85]. 2. *Environmental impact assessment of strategic tillage*: the primary objectives of the assessment are to understand the effects of strategic tillage on various environmental aspects, identify potential hazards, and propose solutions to mitigate negative impacts. The assessment aims to determine the likelihood of soil erosion caused by tillage activities and assess the effectiveness of strategic tillage in reducing erosion [86]. The potential for sediment discharge, leading to water pollution and sludge buildup in water bodies, is also analyzed. The effects of strategic tillage on soil health are investigated, including modifications to soil structure, organic matter content, nutrient levels, and microbial activity. Long-term or heavy tillage techniques can lower soil quality and its ability to support healthy plant growth [87]. The impact of strategic tillage on plant, insect, and soil biodiversity is examined. Reduced tillage practices generally promote biodiversity by providing habitat and minimizing habitat degradation caused by conventional tillage. The assessment looks into the potential release of nutrients and pesticides from tilled fields, which can lead to water pollution. A comparison is made between the effects of strategic tillage and traditional tillage on water quality [88]. The analysis focuses on how strategic tillage affects water use efficiency and irrigation needs. Reduced tillage techniques may help retain soil moisture, thus improving water use efficiency. The assessment includes an analysis of greenhouse gas emissions associated with strategic tillage, such as carbon dioxide (CO_2) and nitrous oxide (N_2O) [89]. Reduced tillage techniques have the potential to sequester carbon in the soil, reducing

emissions. The impact of strategic tillage techniques on weed and pest populations is analyzed. Reduced tillage practices that disturb the soil less may alter pest dynamics. The assessment examines how strategic tillage affects crop yields and overall agricultural productivity, with the goal of supporting or enhancing crop output through sustainable tillage practices [86]. Potential effects on wildlife habitat, environmental esthetics, and changes in land use patterns resulting from strategic tillage are also analyzed in the assessment.

9. Future directions and emerging trends

Modern farming techniques that attempt to increase output while reducing environmental impacts include strategic tillage and sustainable agriculture. The need for creative and sustainable ways is becoming more and more obvious as the agriculture sector deals with issues like climate change, soil degradation, and population expansion [90]. Strategic tillage and sustainable agriculture are being shaped by a number of rising trends, pushing the sector toward more effective and environmentally friendly methods [91]. *Precision farming*: precision farming maximizes the use of resources like water, fertilizer, and pesticides by utilizing cutting-edge technologies like GPS, remote sensing, and data analytics [92]. Farmers can customize their tillage operations to meet the particular requirements of each fields by using site-specific management techniques, which results in higher yields and fewer negative environmental effects [93]. *Conservation tillage*: in conservation tillage, crop wastes are left on the field after harvest with minimal soil disturbance. Long-term sustainability is improved by this approach because it improves soil health, lessens erosion, and sequesters carbon in the soil [94]. This group includes no-till and reduced-till techniques, which are gaining acceptance due to their capacity to preserve soil moisture and enhance general soil structure. *Climate-resilient farming*: farming practices are becoming more climate-resilient as a result of how climate change is affecting weather patterns and escalating the frequency of extreme events [95]. This entails varying crop types, utilizing drought-tolerant cultivars, and putting in place irrigation techniques that use little water. *Agroforestry and agroecology*: agroforestry involves integrating trees with agricultural crops to create sustainable and biodiverse systems [96]. These systems not only provide multiple income streams but also enhance soil fertility, biodiversity, and carbon sequestration. Agroecology, on the other hand, promotes a systems-thinking approach, focusing on enhancing the ecological interactions within the farm ecosystem [97]. *Data-driven decision making*: big data and sophisticated analytics are revolutionizing the agricultural industry. Through real-time data gathering and analysis, tillage practice optimization, and resource efficiency, farmers can now make better decisions [98].

10. Conclusion

Harnessing soil potential through innovation in strategic tillage and management practices is crucial for ensuring sustainable agricultural productivity and environmental health. Throughout this chapter, we have explored various aspects of soil management and how they contribute to optimizing crop yields while minimizing negative environmental impacts. Modern agriculture faces a number of difficulties, including soil deterioration, nutrient depletion, and water shortages, which emphasize the pressing, need to use sustainable soil management techniques. Conventional tillage techniques have shown to be successful in the short term, but they can have negative long-term effects such soil erosion and loss of soil fertility. We can maintain soil structure, hold onto organic matter, and boost soil health by using alternative tillage techniques including conservation tillage, reduced tillage, and no-till farming. However, farmer education and extension services are necessary for the effective adoption of new soil management practices. In order to ensure widespread acceptance and long-term success, it will be essential to equip farmers with the information and abilities to embrace sustainable practices. In addition, encouraging government policies and incentives are required to promote the use of sustainable agriculture methods. Governments may play a key role in promoting sustainable soil management by offering financial incentives, technical assistance, and legislative frameworks. In conclusion, the key to attaining sustainable agriculture is to maximize the potential of the soil via innovation in strategic tillage and management practices. We can make future generations more food secure and resilient by maintaining soil health, maximizing resource utilization, and reducing environmental consequences. Protecting the wellbeing of both people and the environment will depend on adopting sustainable soil management techniques.

Author details

Sajid Ali[1], Adnan Zahid[1*], Ammara Fatima[2], Mukhtar Ahmad[3], Muhammad Tariq Manzoor[4], Asma Ayub[1], Ahmad Raza[1] and Nauman Shafqat[1]

1 Department of Agronomy, University of the Punjab, Lahore, Pakistan

2 Department of Environmental Science, Lahore College for Women University, Lahore, Pakistan

3 Department of Agronomy, Pir Mehr Ali Shah Arid Agriculture University, Rawalpindi, Pakistan

4 Department of Plant Pathology, University of the Punjab, Lahore, Pakistan

*Address all correspondence to: adnan.iags@pu.edu.pk

References

[1] Bagnall DK, Shanahan JF, Flanders A, Morgan CL, Honeycutt CW. Soil health considerations for global food security. Agronomy Journal. 2021;**113**(6):4581-4589

[2] Tilman D, Balzer C, Hill J, Befort BL. Global food demand and the sustainable intensification of agriculture. Proceedings of the National Academy of Sciences. 2011;**108**(50):20260-20264

[3] Wander MM, Cihacek LJ, Coyne M, Drijber RA, Grossman JM, Gutknecht JL, et al. Developments in agricultural soil quality and health: Reflections by the research committee on soil organic matter management. Frontiers in Environmental Science. 2019;**7**:109

[4] Lima A, Brussaard L, Totola M, Hoogmoed W, De Goede R. A functional evaluation of three indicator sets for assessing soil quality. Applied Soil Ecology. 2013;**64**:194-200

[5] Masto RE, Chhonkar PK, Singh D, Patra AK. Soil quality response to long-term nutrient and crop management on a semi-arid Inceptisol. Agriculture, Ecosystems & Environment. 2007;**118**(1-4):130-142

[6] Armenise E, Redmile-Gordon M, Stellacci A, Ciccarese A, Rubino P. Developing a soil quality index to compare soil fitness for agricultural use under different managements in the Mediterranean environment. Soil and Tillage Research. 2013;**130**:91-98

[7] Yuan P, Wang J, Li C, Xiao Q, Liu Q, Sun Z, et al. Soil quality indicators of integrated rice-crayfish farming in the Jianghan Plain, China using a minimum data set. Soil and Tillage Research. 2020;**204**:104732

[8] Almagro M, Garcia-Franco N, Martínez-Mena M. The potential of reducing tillage frequency and incorporating plant residues as a strategy for climate change mitigation in semiarid Mediterranean agroecosystems. Agriculture, Ecosystems & Environment. 2017;**246**:210-220

[9] Biddoccu M, Ferraris S, Pitacco A, Cavallo E. Temporal variability of soil management effects on soil hydrological properties, runoff and erosion at the field scale in a hillslope vineyard, North-West Italy. Soil and Tillage Research. 2017;**165**:46-58

[10] Martínez-Mena M, Perez M, Almagro M, García-Franco N, Díaz-Pereira E. Long-term effects of sustainable management practices on soil properties and crop yields in rainfed Mediterranean almond agroecosystems. European Journal of Agronomy. 2021;**123**:126207

[11] Aguilera E, Lassaletta L, Gattinger A, Gimeno BS. Managing soil carbon for climate change mitigation and adaptation in Mediterranean cropping systems: A meta-analysis. Agriculture, Ecosystems & Environment. 2013;**168**:25-36

[12] Francaviglia R, Almagro M, Vicente-Vicente JL. Conservation agriculture and soil organic carbon: Principles, processes, practices and policy options. Soil Systems. 2023;**7**(1):17

[13] Moraru PI, Rusu T. No-tillage and minimum tillage systems with reduced energy consumption and soil conservation in the hilly areas of Romania. Journal of Food, Agriculture & Environment. 2013;**11**(2):1227-1231

[14] Pardo G, Del Prado A, Martínez-Mena M, Bustamante M,

Martín JR, Álvaro-Fuentes J, et al. Orchard and horticulture systems in Spanish Mediterranean coastal areas: Is there a real possibility to contribute to C sequestration? Agriculture, Ecosystems & Environment. 2017;**238**:153-167

[15] Sheehy J, Regina K, Alakukku L, Six J. Impact of no-till and reduced tillage on aggregation and aggregate-associated carbon in Northern European agroecosystems. Soil and Tillage Research. 2015;**150**:107-113

[16] Jayaraman S, Dalal RC. No-till farming: Prospects, challenges – productivity, soil health, and ecosystem services. CSIRO Publishing; 2022;**60**(6):435-441

[17] Lobb DA, Huffman E, Reicosky DC. Importance of information on tillage practices in the modelling of environmental processes and in the use of environmental indicators. Journal of Environmental Management. 2007;**82**(3):377-387

[18] Ball B, Bingham I, Rees R, Watson C, Litterick A. The role of crop rotations in determining soil structure and crop growth conditions. Canadian Journal of Soil Science. 2005;**85**(5):557-577

[19] Keller T, Colombi T, Ruiz S, Schymanski SJ, Weisskopf P, Koestel J, et al. Soil structure recovery following compaction: Short-term evolution of soil physical properties in a loamy soil. Soil Science Society of America Journal. 2021;**85**(4):1002-1020

[20] Qi H, Paz-Kagan T, Karnieli A, Jin X, Li S. Evaluating calibration methods for predicting soil available nutrients using hyperspectral VNIR data. Soil and Tillage Research. 2018;**175**:267-275

[21] Krauss M, Berner A, Perrochet F, Frei R, Niggli U, Mäder P. Enhanced soil quality with reduced tillage and solid manures in organic farming–A synthesis of 15 years. Scientific Reports. 2020;**10**(1):4403

[22] Acir N, Günal H, Celik I, Barut ZB, Budak M, Kılıç Ş. Effects of long-term conventional and conservational tillage systems on biochemical soil health indicators in the Mediterranean region. Archives of Agronomy and Soil Science. 2022;**68**(6):795-808

[23] Cherry K, Mila A. Temporal progress and control of tomato spotted wilt virus in flue-cured tobacco. Crop Protection. 2011;**30**(5):539-546

[24] Nord EA, Ryan MR, Curran WS, Mortensen DA, Mirsky SB. Effects of management type and timing on weed suppression in soybean no-till planted into rolled-crimped cereal rye. Weed Science. 2012;**60**(4):624-633

[25] Singh D, Ranjan P, Solanki K, Sandeep S. Effect of tillage and weed management practices on dry matter, yield and nutrient uptake by plant and depletion by Weedin Lentil Crop (*Lens culinaris* M.). International Journal of Environment and Climate Change. 2023;**13**(9):288-298

[26] Sarkar S, Skalicky M, Hossain A, Brestic M, Saha S, Garai S, et al. Management of crop residues for improving input use efficiency and agricultural sustainability. Sustainability. 2020;**12**(23):9808

[27] Kaur R, Kaur S, Deol JS, Sharma R, Kaur T, Brar AS, et al. Soil properties and weed dynamics in wheat as affected by rice residue management in the rice–wheat cropping system in south Asia: A review. Plants. 2021;**10**(5):953

[28] Zhang Y, Xie D, Ni J, Zeng X. Optimizing phosphate fertilizer application to reduce nutrient loss in a

mustard (*Brassica juncea* var. *tumida*)-maize (*Zea mays* L.) rotation system in three gorges reservoir area. Soil and Tillage Research. 2019;**190**:78-85

[29] Basset C, Abou Najm M, Ghezzehei T, Hao X, Daccache A. How does soil structure affect water infiltration? A meta-data systematic review. Soil and Tillage Research. 2023;**226**:105577

[30] Akbarnia A, Farhani F, Heidary B. Economic comparison of tillage and planting operations in three tillage systems. Agricultural Engineering International: CIGR Journal. 2013;**15**(4):180-184

[31] Mia MS, Azam G, Nouraei S, Borger C. Strategic tillage in Australian conservation agricultural systems to address soil constraints: How does it impact weed management? Weed Research. 2023;**63**(1):12-26

[32] Oreja FH, Batlla D, de la Fuente EB. *Digitaria sanguinalis* seed dormancy release and seedling emergence are affected by crop canopy and stubble. Weed Research. 2020;**60**(2):111-120

[33] Schomberg HH, McDaniel RG, Mallard E, Endale DM, Fisher DS, Cabrera ML. Conservation tillage and cover crop influences on cotton production on a southeastern US coastal plain soil. Agronomy Journal. 2006;**98**(5):1247-1256

[34] Zeng Z, Thoms D, Chen Y, Ma X. Comparison of soil and corn residue cutting performance of different discs used for vertical tillage. Scientific Reports. 2021;**11**(1):2537

[35] Gupta PR, Chandan YN, Banginwar AR, Gawande GR, Inglekar PS, Shrirao PN. An overview on design analysis and development of agricultural rotavator blade. International Journal of Advance Scientific Research and Engineering Trends. 2021;**6**(4):35-36

[36] Rizwan M, Rubina Gilani S, Iqbal Durani A, Naseem S. Materials diversity of hydrogel: Synthesis, polymerization process and soil conditioning properties in agricultural field. Journal of Advanced Research. 2021;**33**:15-40

[37] Chunlei W, Hongwen L, Jin H, Qingjie W, Caiyun L, Liping C. State-of-the-art and Prospect of automatic navigation and measurement techniques application in conservation tillage. Smart Agriculture. 2020;**2**(4):41

[38] McLaughlin N, Drury C, Reynolds W, Yang X, Li Y, Welacky T, et al. Energy inputs for conservation and conventional primary tillage implements in a clay loam soil. Transactions of the ASABE. 2008;**51**(4):1153-1163

[39] Tessier S, Hyde G, Papendick R, Saxton K. No-till seeders effects on seed zone properties and wheat emergence. Transactions of the ASAE. 1991;**34**(3):733-0739

[40] Noland RL, Wells MS, Sheaffer CC, Baker JM, Martinson KL, Coulter JA. Establishment and function of cover crops interseeded into corn. Crop Science. 2018;**58**(2):863-873

[41] Lin J, Liu A, Li B, Li B, Zhao D, Lu C. 2BG-2 type corn ridge planting no-till planter. Nongye Jixie Xuebao = transactions of the Chinese Society for Agricultural Machinery. 2011;**42**(6):43-62

[42] Somasundaram J, Sinha N, Dalal RC, Lal R, Mohanty M, Naorem A, et al. No-till farming and conservation agriculture in South Asia–issues, challenges, prospects and benefits. Critical Reviews in Plant Sciences. 2020;**39**(3):236-279

[43] Adetunji AT, Ncube B, Mulidzi R, FB L. Management impact and benefit of cover crops on soil quality: A review, Soil and Tillage Research. 2020;**204**:104717

[44] Gómez JA, Campos M, Guzmán G, Castillo-Llanque F, Vanwalleghem T, Lora Á, et al. Soil erosion control, plant diversity, and arthropod communities under heterogeneous cover crops in an olive orchard. Environmental Science and Pollution Research. 2018;**25**:977-989

[45] Delgado JA, Mosquera VHB, Alwang JR, Villacis-Aveiga A, Ayala YEC, Neer D, et al. Potential use of cover crops for soil and water conservation, nutrient management, and climate change adaptation across the tropics. Advances in Agronomy. 2021;**165**:175-247

[46] López-Vicente M, Calvo-Seas E, Álvarez S, Cerdà AJ. Effectiveness of cover crops to reduce loss of soil organic matter in a rainfed vineyard. Land. 2020;**9**(7):230

[47] Seifert CA, Azzari G, Lobell DB. Satellite detection of cover crops and their effects on crop yield in the Midwestern United States., Environmental Research Letters. 2018;**13**(6):064033

[48] Kim N, Zabaloy MC, Guan K, Villamil MBJ. Do cover crops benefit soil microbiome? A meta-analysis of current research. Soil Biology and Biochemistry. 2020;**142**:107701

[49] de Pedro L, Perera-Fernández LG, López-Gallego E, Pérez-Marcos M, Sanchez JA. The effect of cover crops on the biodiversity and abundance of ground-dwelling arthropods in a Mediterranean pear orchard. Agronomy. 2020;**10**(4):580

[50] Meyer N, Bergez J-E, Constantin J, Belleville P, Justes E. Cover crops reduce drainage but not always soil water content due to interactions between rainfall distribution and management. Agricultural Water Management. 2020;**231**:105998

[51] Kaye JP, Quemada M. Using cover crops to mitigate and adapt to climate change. A review. Agronomy for Sustainable Development. 2017;**37**:1-17

[52] Chapagain T, Lee EA, Raizada MNJ. The potential of multi-species mixtures to diversify cover crop benefits. Sustainability. 2020;**12**(5):2058

[53] Wu H, Lai C, Zeng G, Liang J, Chen J, Xu J, et al. The interactions of composting and biochar and their implications for soil amendment and pollution remediation: A review. Critical Reviews in Biotechnology. 2017;**37**(6):754-764

[54] Alley MM, Vanlauwe B. The Role of Fertilizers in Integrated Plant Nutrient Management. United Kingdom: International Fertilizer Industry Association; 2009

[55] Tully K, Ryals R. Nutrient cycling in agroecosystems: Balancing food and environmental objectives. Agroecology and Sustainable Food Systems. 2017;**41**(7):761-798

[56] Agegnehu G, Amede T, Erkossa T, Yirga C, Henry C, Tyler R, et al. Extent and management of acid soils for sustainable crop production system in the tropical agroecosystems: A review. Acta Agriculturae Scandinavica, Section B — Soil & Plant Science. 2021;**71**(9):852-869

[57] Neina DJ. The role of soil pH in plant nutrition and soil remediation. Applied and Environmental Soil Science. 2019;**2019**:1-9

[58] Frouz JJ. Effects of soil macro-and mesofauna on litter decomposition

and soil organic matter stabilization. Geoderma. 2018;**332**:161-172

[59] Vecchio Y, De Rosa M, Adinolfi F, Bartoli L, Masi M. Adoption of precision farming tools: A context-related analysis. Land Use Policy. 2020;**94**:104481

[60] Radoglou-Grammatikis P, Sarigiannidis P, Lagkas T, Moscholios IJ. A compilation of UAV applications for precision agriculture. Computer Networks. 2020;**172**:107148

[61] Rwanga SS, Ndambuki JM. Accuracy assessment of land use/ land cover classification using remote sensing and GIS. International Journal of Geosciences. 2017;**8**(04):611

[62] Singh T, Asim M. Weather monitoring system using IoT. In: Innovations in Cyber Physical Systems: Select Proceedings of ICICPS 2020. Germany: Springer; 2021

[63] Späti K, Huber R, Finger RJEE. Benefits of increasing information accuracy in variable rate technologies. Ecological Economics. 2021;**185**:107047

[64] Dolci R, editor. IoT solutions for precision farming and food manufacturing: Artificial intelligence applications in digital food. In: 2017 IEEE 41st Annual Computer Software and Applications Conference (COMPSAC). Canada: IEEE; 2017

[65] Phupattanasilp P, Tong S-R. Augmented reality in the integrative internet of things (AR-IoT): Application for precision farming. Sustainability. 2019;**11**(9):2658

[66] AJT G, Fullen MA, Jorge MCO, JFR B, Shokr MS. Slope processes, mass movement and soil erosion: A review. Pedosphere. 2017;**27**(1):27-41

[67] Merrison J. Sand transport, erosion and granular electrification. Aeolian Research. 2012;**4**:1-16

[68] Van Oost K, Govers G, De Alba S, Quine T. Tillage erosion: A review of controlling factors and implications for soil quality. Progress in Physical Geography. 2006;**30**(4):443-466

[69] Ruysschaert G, Poesen J, Verstraeten G, Govers G. Soil loss due to crop harvesting: Significance and determining factors. Progress in Physical Geography. 2004;**28**(4):467-501

[70] Kuhwald M, Busche F, Saggau P, Duttmann R. Is soil loss due to crop harvesting the most disregarded soil erosion process? A review of harvest erosion. Soil and Tillage Research. 2022;**215**:105213

[71] Borrelli P, Robinson DA, Fleischer LR, Lugato E, Ballabio C, Alewell C, et al. An assessment of the global impact of 21st century land use change on soil erosion. Nature Communications. 2017;**8**(1):2013

[72] Zachar D. Soil erosion. United Kingdom: Elsevier; 2011

[73] Fayas CM, Abeysingha NS, Nirmanee KGS, Samaratunga D, Mallawatantri A. Soil loss estimation using rusle model to prioritize erosion control in KELANI river basin in Sri Lanka. International Soil and Water Conservation Research. 2019;**7**(2):130-137

[74] Lal R. Soil erosion impact on agronomic productivity and environment quality. Critical Reviews in Plant Sciences. 1998;**17**(4):319-464

[75] Lal R. Soil degradation by erosion. Land Degradation & Development. 2001;**12**(6):519-539

[76] Bandumula N. Rice production in Asia: Key to global food security. Proceedings of the National Academy of Sciences, India Section B: Biological Sciences. 2018;**88**:1323-1328

[77] Wang L, Yuan X, Liu C, Li Z, Chen F, Li S, et al. Soil C and N dynamics and hydrological processes in a maize-wheat rotation field subjected to different tillage and straw management practices. Agriculture, Ecosystems & Environment. 2019;**285**:106616

[78] Pan D, Zhao X, Gao X, Song Y, Dyck M, Wu P, et al. Application rate influences the soil and water conservation effectiveness of mulching with chipped branches. Soil Science Society of America Journal. 2018;**82**(2):447-454

[79] Dai C, Liu Y, Wang T, Li Z, Zhou Y. Exploring optimal measures to reduce soil erosion and nutrient losses in southern China. Agricultural Water Management. 2018;**210**:41-48

[80] Sharma N, Singh RJ, Mandal D, Kumar A, Alam N, Keesstra S. Increasing farmer's income and reducing soil erosion using intercropping in rainfed maize-wheat rotation of Himalaya, India. Agriculture, Ecosystems & Environment. 2017;**247**:43-53

[81] Keesstra S, Rodrigo-Comino J, Novara A, Giménez-Morera A, Pulido M, Di Prima S, et al. Straw mulch as a sustainable solution to decrease runoff and erosion in glyphosate-treated clementine plantations in eastern Spain. An assessment using rainfall simulation experiments. Catena. 2019;**174**:95-103

[82] Cerdà A, Rodrigo-Comino J, Giménez-Morera A, Keesstra SD. Hydrological and erosional impact and farmer's perception on catch crops and weeds in citrus organic farming in Canyoles river watershed, Eastern Spain. Agriculture, Ecosystems & Environment. 2018;**258**:49-58

[83] Ahmad NSBN, Mustafa FB, Didams G. A systematic review of soil erosion control practices on the agricultural land in Asia. International Soil and Water Conservation Research. 2020;**8**(2):103-115

[84] Azumah SB, Adzawla W, Osman A, Anani PY. Cost-benefit analysis of on-farm climate change adaptation strategies in Ghana. Ghana Journal of Geography. 2020;**12**(1):29-46

[85] Shongwe P. Cost Benefit Analysis of Climate Change Adaption Strategies on Crop Production Systems: A Case of Mpolonjeni Area Development Programme (ADP) in Swaziland. 2013

[86] Wang X, Qi JY, Liu BY, Kan ZR, Zhao X, Xiao XP, et al. Strategic tillage effects on soil properties and agricultural productivity in the paddies of southern China. Land Degradation & Development. 2020;**31**(10):1277-1286

[87] Brentrup F, Küsters J, Lammel J, Barraclough P, Kuhlmann H. Environmental impact assessment of agricultural production systems using the life cycle assessment (LCA) methodology II. The application to N fertilizer use in winter wheat production systems. European Journal of Agronomy. 2004;**20**(3):265-279

[88] Dang Y, Seymour NP, Walker S, Bell M, Freebairn D. Strategic tillage in no-till farming systems in Australia's northern grains-growing regions: I. Drivers and implementation. Soil and Tillage Research. 2015;**152**:104-114

[89] Melland A, Antille D, Dang Y. Effects of strategic tillage on short-term erosion, nutrient loss in runoff and

greenhouse gas emissions. Soil Research. 2016;**55**(3):201-214

[90] Guimarães RM, Lamandé M, Munkholm LJ, Ball BC, Keller T. Opportunities and future directions for visual soil evaluation methods in soil structure research. Soil and Tillage Research. 2017;**173**:104-113

[91] Dang YP, Dalal RC, Menzies NW. No-Till Farming Systems for Sustainable Agriculture: Challenges and Opportunities. Sweden: Springer; 2020

[92] Blasch J, van der Kroon B, van Beukering P, Munster R, Fabiani S, Nino P, et al. Farmer preferences for adopting precision farming technologies: A case study from Italy. European Review of Agricultural Economics. 2022;**49**(1):33-81

[93] Raj EFI, Appadurai M, Athiappan K. Precision farming in modern agriculture. In: Smart Agriculture Automation Using Advanced Technologies: Data Analytics and Machine Learning, Cloud Architecture, Automation and IoT. Switzerland: Springer; 2022. pp. 61-87

[94] Hussain S, Hussain S, Guo R, Sarwar M, Ren X, Krstic D, et al. Carbon sequestration to avoid soil degradation: A review on the role of conservation tillage. Plants. 2021;**10**(10):2001

[95] Amoak D, Luginaah I, McBean G. Climate change, food security, and health: Harnessing agroecology to build climate-resilient communities. Sustainability. 2022;**14**(21):13954

[96] Buratti-Donham J, Venn R, Schmutz U, Migliorini P. Transforming food systems towards agroecology – A critical analysis of agroforestry and mixed farming policy in 19 European countries. Agroecology and Sustainable Food Systems. 2023;**47**(7):1-29

[97] Peredo Parada S, Barrera C, Burbi S, Rocha D. Agroforestry in the Andean Araucanía: An experience of agroecological transition with women from Cherquén in Southern Chile. Sustainability. 2020;**12**(24):10401

[98] Kamble SS, Gunasekaran A, Gawankar SA. Achieving sustainable performance in a data-driven agriculture supply chain: A review for research and applications. International Journal of Production Economics. 2020;**219**:179-194